Electricity's Future: The Shift to Efficiency And Small-Scale Power

Christopher Flavin

Worldwatch Paper 61
November 1984

Library of Congress Catalog Card Number 84-52357
ISBN 0-916468-61-5

Printed on recycled paper

Table of Contents

Electric power has largely supplanted oil as the most controversial energy issue of the eighties. Soaring costs, high interest rates, and environmental damage caused by large power plants have wreaked havoc on the once booming electricity industry. In most countries, electric prices have risen faster than the general rate of inflation since the mid-seventies. Nuclear reactors, once expected to be the main source of power in the eighties and beyond, have been plagued by technical breakdowns and staggering cost overruns.

Electricity's future role is more uncertain than at any time since Thomas Edison opened the world's first commercial power plants in Manhattan and London in 1882. Policymakers around the world disagree vigorously about future trends and appropriate policies, but virtually all acknowledge that a turning point has been reached. The world is unlikely to return to the steady, predictable growth of the past. New solutions will be required to put power generation on an environmentally sound and economically sustainable footing.

Electricity is now at the center of some of the world's bitterest economic and environmental controversies. Nuclear energy, which until recently was viewed as the only new electricity source the world would need in the eighties, has suffered a series of setbacks, undermining public confidence and driving up costs. So far the problems have defied quick-fix solutions, and projections of nuclear power's future contribution have shrunk drastically.

Coal also faces serious hurdles. Coal-fired power plants are a major cause of air pollution, and are implicated in the predominant environmental issue of the eighties: acid rain. There is growing evidence that acid rain is damaging the world's forests. Heavy reliance on coal

I would like to thank Cynthia Pollock for research assistance in preparing this publication.

might require the sacrifice of some major forests, and this has sparked efforts to limit coal-related air pollution. Technologies are being developed for cleaner coal combustion, but the uncertainty and cost of these solutions clouds the future of coal as a power source.

Further complicating the picture is the slower and more erratic growth in power use during the past few years. Electricity forecasts made a decade ago by industry and government overestimated 1984 consumption by an amount equal to the output of several hundred nuclear plants costing hundreds of billions of dollars. The reason: higher prices have encouraged electricity conservation. Nonetheless, electricity as a share of total energy use has risen during the past decade, replacing oil in some applications.

Many in the power business maintain that new demands will keep electricity use growing faster than most other sectors of the world economy. But the more efficient appliances and industrial equipment now available make this unlikely. In fact, rapid introduction of such technologies, spurred by higher electricity prices, might lower power use regardless of the rapid growth and "electrification" of some parts of the economy.

The remarkably divergent predictions of different forecasters have led, not surprisingly, to a planning paralysis among utility executives and government regulators. Day-to-day business is now dominated by arguments over forecast accuracy, who will pay for unnecessary plants, and how rapidly electricity prices should be permitted to increase. While the muddle-through strategy prevails, many signs indicate that it is not working.

Amid the confusion and hand wringing, many planners have missed the most important development in the early eighties: large central power plants no longer entirely dominate electricity planning. Since 1980, cancellations of nuclear and coal plants in the United States have far outrun new orders. In other countries plant orders have slowed to a trickle. Meanwhile, 785 small-scale power projects, with a

"A few utilities, particularly those in California, have encouraged small-scale power production, with impressive results."

total generating capacity of 14,000 megawatts, have been registered with the U.S. Federal Energy Regulatory Commission. Most will begin generating power within a few years. These projects will provide enough power to supply 4 million homes, or to satisfy two years of growth in U.S. power demand. The new sources include a mix of cogeneration, biomass, small hydropower, wind power, and geothermal energy.

This rush to small-scale power production is not being led by utilities. Leading the way instead are large industrial companies building their own power systems and small firms created to tap new energy sources such as wind power and geothermal energy. Utilities buy power from the "small producers" and distribute it to customers. Behind much of this activity is legislation passed in the late seventies and court rulings in the early eighties that have ended utilities' monopoly control of power generation in the United States. Federal and state tax incentives have also encouraged development of some of the new technologies. The resulting boom in small-scale power production is a good example of what can happen when rapid advances in technology are joined by entrepreneurial capitalism. The cost of the new power sources is falling steadily. Some are already less expensive than recent coal and nuclear plants, and others soon will be.

The blossoming of small-scale power generation has been largely ignored or actively obstructed by the utility industry. The Edison Electric Institute, an association of private U.S. utilities, excludes new energy sources from its power generation statistics and assumes that future energy needs will be met by large coal and nuclear plants. Many utilities offer only 2-4¢ per kilowatt-hour for this power while spending over 10¢ per kilowatt-hour to harness power on their own. However, a few utilities, particularly those in California, have encouraged small-scale power production, with impressive results. Based on recent trends, California may get 20 percent of its power from these energy sources by 1990. In California and elsewhere, encouragement from state regulatory commissions has been a prerequisite to such a shift.

8 Because most countries have rigid, centralized utility systems, small-scale power generation has barely caught on outside the United States. In many countries a single state utility or a few large private utilities have exclusive rights to generate power, and these bureaucracies have concentrated on large power plants. But rapid advances under way in a wide range of small-scale generating technologies may soon encourage changes worldwide. Research programs are widespread, and international developments are closely followed.

Improved energy efficiency and load management should also be considered as alternatives to building new power plants. In most regions of the world inefficient appliances can be replaced, houses weatherized, and industrial equipment upgraded for a fraction of the cost of building a new generating plant. Efficiency can be promoted many ways, but some of the best include utility-sponsored information and financing programs, with a return allowed on the investment, just as a new power plant would receive. Electricity prices can be adjusted to encourage less power use at peak periods, thus avoiding the need to build additional plants. Many utilities have recently adopted efficiency programs at the insistence of government regulators, but most are just token efforts.

The utility industry as a whole has become staid and lethargic. Executives in both public and private companies tend to view their business as building enough capacity to meet predicted energy demand. Success is measured by how well they carry out this service. Government regulators generally limit themselves to approving construction programs and granting the revenues needed for utilities to effectively serve their customers. No one takes responsibility for asking fundamental questions or challenging accepted practices. Lack of competition has clearly taken its toll.

The electricity business is in need of fundamental structural change. Utility monopoly of power generation hinders research on new technologies and development of small-scale electricity sources. Advances in energy efficiency have been slowed because investments in

> **"Stung by rate increases, many ordinary citizens have become involved in utility affairs for the first time."**

efficiency rarely get the subsidies and tax breaks that many countries give to new power plants. The electricity industry's roots lie in the efforts of Thomas Edison and other early inventors and entrepreneurs. A similar spirit of innovation is badly needed today. In short, the utility industry must become more competitive and market-oriented, while still providing the reliable, affordable service it is known for.

Major changes in institutions rarely happen unless prompted by a crisis. In the United States, the utility industry's economic and environmental problems of the past five years have already led to innovations as significant as any in the history of the power business. Regulators in some states have effectively broken the monopoly on power generation once held by utilities. Stung by rate increases, many ordinary citizens have become involved in utility affairs for the first time. They have demanded that utility incentives be restructured to reward high performance and penalize companies making faulty investments with ratepayers' money. Where these changes will ultimately lead is hard to say, but the electricity business a decade hence will surely be quite different than it is now.

The End of an Era

A century ago, only a few wealthy neighborhoods in the largest cities of Europe and America enjoyed the use of electric power. Yet by 1900, electricity was already essential to modern industry and society, lighting homes and powering public transportation systems. Tens of thousands of power plants sprang up to meet electricity needs, and the United States alone had over 4,000 utility companies. Although small by modern standards, the coal-fired and hydropower plants built in 1900 were close cousins of those in use today.[1]

During the early years of the century, the electricity business attracted ambitious entrepreneurs, and electricity use grew rapidly. Some cities were served by as many as a dozen competing utilities that would run power lines down opposite sides of the same street. In addition, many industries generated their own power, independent of the burgeoning utility industry. Adding further diversity were municipally owned utilities, often established after intense political battles. From the trolley cars of the 1890's to the microelectronics of the 1970's, growth in electric power use paralleled growth of the world economy, pausing only briefly during such epic twentieth century events as the Great Depression and the two world wars.[2]

The economies of scale of large steam turbines, the falling costs of electricity transmission, and the free market spirit generally favored larger, privately owned utilities during electricity's early years. Throughout the United States, small generating stations were shut down during the teens and twenties, and the electricity business gradually became a monopoly. Between 1917 and 1927, 900 municipal utilities went out of business. By the early thirties George Insull, president of Commonwealth Edison, controlled 65 private power companies.[3]

Concerned that this important industry had become monopolized, the U.S. Congress passed the Public Utility Holding Company Act of 1935, breaking up some of the vast electric conglomerates but leaving many large investor-owned utilities intact. Another New Deal law created the Rural Electrification Administration to provide low-interest, government-backed loans for electric cooperatives that would provide power for farms. The rural electric cooperative movement led eventually to over 1,000 co-ops that own 42 percent of the country's distribution lines and serve 25 million people. In the Southeast a vast program of dam building and electrification began in the thirties with the federally owned Tennessee Valley Authority, today the nation's largest electric utility. Only a tenth of U.S. farms had electricity in 1930, but 43 percent had power in 1943 and 98 percent in 1975.[4]

Despite extensive publicly supported electrification programs, the power business in the United States remains largely in private hands.

About 200 investor-owned utilities serve three-quarters of America's households and generate four-fifths of the country's electricity. Another 10 percent is generated at federally owned hydro projects and sold to public and private utilities. About 2,200 public utilities, most of them municipally owned, generate less than 10 percent of the country's power. Like the electric co-ops, municipals buy most of the electricity they distribute from private utilities.

The electrification of Europe matched progress in the United States in the early part of the century and included a complex mix of public and private utilities. In Great Britain, 572 private and municipal utilities operated 491 power stations by 1925. A privately owned utility industry dominated in France, Italy, and Germany, though some public utilities also operated, providing, for example, one-quarter of Italy's power.[5]

In the Soviet Union, where electrification had hardly begun before the 1917 Bolshevik Revolution, state ownership of the electric power system was firmly established in 1920 under the State Commission for the Electrification of Russia. Since then, electrification of Soviet industry has been a high priority in the country's Five-Year Plans. Hundreds of large hydro and thermal generating stations and one of the world's most extensive transmission systems have been built. The main Soviet power grid now includes 800,000 kilometers of transmission lines that serve an area of over 10 million square kilometers.[6]

During the postwar period, electricity use grew at a rate of 5 to 10 percent per year in most industrial countries, and this fast pace put enormous strains on often fragmented supply systems. As a result, many countries chose to nationalize their utilities. In Great Britain, 95 percent of the country's electric industry was taken over by the government in 1948, and today the Central Electricity Generating Board owns most of the power plants and transmission lines, while 12 regional boards are responsible for distribution. Electricité de France took over France's electricity system the same year. It is now the world's largest utility and has the biggest nuclear construction program.[7]

In West Germany, the country's approximately 4,000 utility companies are privately owned, but most are partly financed by either state or federal governments. A network of bureaucrats, industrialists, and bankers conducts most of the planning. Japan's private utilities have presided over one of the most rapid electrification programs in history. From 1950 to 1970 electricity generation in Japan increased sevenfold. The country's utility industry is dominated by nine large investor-owned utilities, led by Tokyo Electric. They work closely with the national government in making plans and allocating investments.[8]

Only in the past 20 years have most developing countries provided electricity for any but their wealthiest citizens. Even today the World Bank reports that less than a quarter of Third World households have a regular supply of electricity, which means that more than two billion people worldwide live without power. In Africa, per capita electricity use is typically one-twentieth of that in Europe. Nonetheless, electricity projects are often given priority by government planners. Industry uses the largest share of the power in developing countries, but as per capita incomes rise and modern office and apartment buildings spring up in cities, electricity growth accelerates. Despite economic problems and higher costs, the World Bank projects that electricity use in developing countries will grow at 6 to 7 percent per year during the next decade, only a little slower than in the seventies.[9]

Overall, hydropower supplies 38 percent of Third World power, oil and gas 31 percent, and coal 30 percent. Industrial countries, by contrast, rely more heavily on coal and nuclear power. In developing countries the resource mix varies widely. Costa Rica and Ghana get virtually all of their electricity from hydropower, while Malaysia and Tunisia get more than three-quarters of their electricity from petroleum. India gets half of its electricity from coal, and China 64 percent.

"Less than a quarter of Third World households have a regular supply of electricity, which means that more than two billion people worldwide live without power."

Wherever possible, developing countries are moving away from oil-fired electricity and building mainly coal and hydro plants.[10]

Electricity prices in developing countries are often higher than in industrial countries, ranging from 4¢ to 20¢ per kilowatt-hour. Industrial users in developing countries pay as much as 50¢ per kilowatt-hour for backup electricity. Providing reliable and economical power is often hampered by the general inefficiency of electricity distribution. Typically, 15 percent of the power generated is "lost," or about twice the normal rate in developed countries. Blackouts occur frequently on the thinly stretched and inadequately maintained power systems of developing countries, and often hamper industrial production.[11]

Third World nations now spend over $40 billion each year on electricity projects, making them the third largest investment after agriculture and transportation. A portion of these funds is supplied by loans and grants from international aid agencies. The World Bank loaned $18.7 billion for 413 electric power projects between 1948 and 1982, and since 1982 has lent $2-$3 billion per year for such projects. Electric power development has been a World Bank priority since the sixties, when it absorbed more than a quarter of total lending. That figure has fallen to about 17 percent in recent years. Electric utilities in developing countries are virtually all government-owned, but because nationwide grids are still rare, countries often have dozens of largely separate power systems.[12]

The fifties and sixties were a time of rapid economic growth and seemingly infinite horizons for electricity in many parts of the world. Power generation and transmission technologies were nearing maturity, and the ever-larger, more advanced plants produced less expensive power than did their predecessors. The average thermal efficiency of U.S. fossil fuel–fired generating plants rose from about 20 percent in the forties to over 30 percent in the sixties. Combined with falling costs of oil and coal, this led to steady declines in U.S. electricity prices throughout the postwar period—from 10.5¢ per kilowatt-hour in 1948 to 2¢ per kilowatt-hour in the sixties.[13]

As electricity became an ever better bargain, its uses grew apace. Many factories were designed to take advantage of the unique properties of electricity, using it to manufacture chemicals, run motors, and perform dozens of other tasks. But residential use of electricity grew even faster, more than doubling during the sixties as refrigerators, dishwashers, air-conditioners, and other appliances became common household amenities. The vast expansion of the "service economy" that began in the sixties also stimulated electricity use. Simply lighting and air-conditioning thousands of fast food restaurants, shopping malls, and offices created a need for many new power plants. High-rise office and apartment buildings that depend on huge air circulation systems to heat, cool, and ventilate provided another boost. Commercial use of electricity in the United States almost tripled between 1960 and the early seventies.[14]

Beyond the versatility and convenience provided by electricity, its use was encouraged by public policies. In most countries, utilities can borrow funds for plant construction at a substantially lower interest rate than do most businesses, often directly through the national treasury. In the United States, public utilities pay no income taxes and can raise money via government-backed, tax-free bonds. U.S. investor-owned utilities have investment tax credits and liberalized depreciation that reduce their effective tax rate to 7.5 percent. Overall, the U.S. utility industry gets close to $10 billion in tax subsidies each year.[15] The income of utilities, determined by state regulators, is generally based on a standard "rate of return" applied to whatever sums utilities invest in new facilities. The best way to increase profits under such an arrangement is to build more plants.

Selling ever greater quantities of electricity and completing ever larger power projects became the key measures of success in the utility business. Advertising and incentive rates were used by utilities and appliance manufacturers to encourage electricity use by their cus-

"By the early seventies planners could no longer assume that each new plant would be more economical than its predecessors."

tomers. With new power plants invariably costing less than old ones, fossil fuel prices declining, and environmental constraints not yet recognized, planners assumed this growth-oriented system would lead to less expensive electricity. During the late sixties, growth in power use in several countries outstripped the bullish pace of plant construction. Although no serious problems resulted, the close call stimulated a new wave of plant building, this time largely nuclear.[16]

The oil price hikes of the seventies had mixed effects on the electricity business. As oil became more expensive, electricity became more attractive when the two competed directly in uses such as home heating and some industrial processes. But many power plants in the early seventies were fueled by oil, so generating costs increased. In the decade following the 1973 oil embargo, the average price paid for fuel oil by U.S. utilities rose more than fivefold. These increases were eventually echoed by a fourfold rise in coal prices and a tenfold increase in natural gas prices. In 1980 alone the fuel costs of U.S. utilities increased by $30 billion, or ten times the utilities' total fuel bills in the early seventies.[17] Because fuel typically accounts for three-quarters of the cost of power generation at an oil-fired plant and half at a coal-fired plant, electricity prices rose immediately. Particularly hurt were consumers in the Far East, much of Europe, Latin America, and some parts of the United States, including the Northeast and West Coast.

Accompanying the rise in fossil fuel prices was a far less publicized loss of technological momentum in electricity generation. By the early seventies planners could no longer assume that each new plant would be more economical than its predecessors. The average thermal efficiency of U.S. fossil fuel–fired plants leveled off, meaning that greater amounts of power could no longer be squeezed out of a given quantity of fuel.[18] Economies of scale in power generation also became more elusive in the seventies. Many large plants had higher generating costs than did smaller plants because the complexity of the plants began to outweigh projected savings. As a result, the average size of new power plants plateaued during the seventies.

Much has been learned about the negative environmental effects of power generation in recent years. Power plants contribute to air and water pollution, they often occupy valuable land, and they can disrupt local communities. In the United States, power plants release 64 percent of sulfur dioxide and 30 percent of nitrogen oxides, pollutants that contribute to respiratory illnesses and forest and crop damage.[19] Environmental opposition to coal and nuclear power plants has sprung up in many parts of the world. In West Germany, political debates and public demonstrations have slowed governmental efforts to gain approval for nuclear plants. In California, the legislature, worried about air quality and radioactive waste, has made it virtually impossible for a utility to gain approval for coal or nuclear plants.

The need for strict pollution controls has raised the cost of electricity generation. A study by energy analyst Charles Komanoff shows that between 1971 and 1978, the real cost of a new coal-fired power plant rose 68 percent, and that most of the increase was caused by pollution control technologies.[20] The most expensive of these is flue gas desulfurization, also called "scrubbing," which has been required in all new coal plants in the United States since 1979. The scrubbers reduce emissions by 70 to 90 percent. The U.S. Environmental Protection Agency estimates that this technology adds 1-1.7¢ per kilowatt-hour to the cost of coal-fired power generation, or a 20 to 40 percent boost for a typical new plant. In 1983, U.S. utilities spent $2.2 billion on pollution control equipment.[21]

In Europe, scrubbers are not generally required, and utilities burn lower sulfur coal and use tall stacks to disperse pollutants. But mounting evidence indicates that acid rain, caused in part by sulfur and nitrogen oxide emissions, is damaging lakes, forests, and cropland. Pressure is building for continent-wide emission reductions under the auspices of the European Economic Community (EEC). Although EEC action is blocked for now by the British government, acid rain controls and further substantial increases in the cost of coal-fired generation are virtually inevitable in most countries.[22] Both in Europe and North America, pressure is growing to retrofit older power plants with scrubbers. Rising carbon dioxide concentrations in the atmosphere, caused by coal and other fossil fuel combustion, are another threat. Many climatologists expect carbon dioxide to cause major changes in the world's climate in the next few decades. So far the only practical strategy to limit carbon dioxide increases is to reduce coal consumption.

The failed promise of nuclear power further complicates efforts to plan electricity's future. As recently as 1970, nuclear plants were expected to provide most of the world's new generating capacity in the nineties. That year, the Organisation for Economic Co-operation and Development (OECD) projected that its member nations in Western Europe, North America, and Japan would have 568,000 megawatts of nuclear capacity by 1985. In reality the total is unlikely to exceed 180,000 megawatts, and orders for additional nuclear plants have slowed to a trickle. In the United States, 110 nuclear projects have been canceled since the mid-seventies. While 125 U.S. nuclear plants were under construction in 1979, cancellations and completions reduced the total to 40 in 1984. Nuclear power's share of electricity generation now ranges from 50 percent in France to 17 percent in Japan, 13 percent in the United States, 6 percent in the Soviet Union, and zero in many industrial and developing nations.[23]

Behind nuclear power's fading fortunes lie problems ranging from the purely technical to the overtly political. Most experts agree that nuclear technology has simply not matured as rapidly as expected, nor have the plants operated as smoothly as had been hoped. Cost trends in the nuclear industry are one indication of the problems that have occurred. Nuclear construction costs during the seventies rose at a real annual rate of 11 percent in Japan, 9 percent in West Germany, 6 percent in Canada, and 5 percent in France. As a result, nuclear costs in most countries barely held their own relative to coal costs and in some cases fell substantially behind. In the United States, generating costs for new nuclear plants went from being lower than those for coal to 65 percent higher.[24]

18 Without a substantially new nuclear technology and a major reordering of the electricity business, nuclear power will provide a limited portion of most countries' electricity in the forseeable future. Exceptions include France, which will get 80 to 90 percent of its electricity from nuclear power by 1990, and Japan, which is scheduled to reach 50 percent in the nineties. In other countries public opposition combined with economic risks will greatly slow development.

Nuclear power has had profound effects on the business of electricity generation. It has played a major role in the financial deterioration of many U.S. utilities and has graphically illustrated the fallacies on which some planning scenarios were based. Nuclear power has prompted a rethinking of such issues as relative risks, the appropriate scale of projects, and dealing with uncertainties. It has also sparked concern from a much wider group of organizations and individuals, making the future of electricity an important social and political question.

Rising nuclear costs, fossil fuel prices, and interest rates naturally led to higher electricity prices. In the United States, the average residential price of electricity rose from 2.5¢ per kilowatt-hour in 1973 to 7.1¢ in 1984. This is a real annual rate of increase of 5.5 percent, slower than price increases for gasoline or natural gas, but substantial nonetheless. Price increases were similar in Europe and even higher in Japan, where dependence on oil-fired generation has pushed prices to 12-15¢ per kilowatt-hour.[25]

Average electricity prices are misleading, however, since they mask enormous variation among regions. In the United States, prices range from 2¢ per kilowatt-hour in Seattle to 8¢ in Kansas City, and 17¢ in New York. Prices are lowest where hydropower is the main energy source and highest where oil is dominant. The very expensive nuclear plants being completed in areas such as Long Island, Arizona, and Texas are expected to cause a doubling of electricity prices. Rising electric rates are a particular burden for low-income consumers and energy-intensive industries. The aluminum industry, for example, is

"Growth in the per capita use of electricity has slowed dramatically in most countries in the past decade."

gradually migrating toward areas of the world with low electricity prices, an option that few individual consumers have.[26]

As prices have risen, the use of electricity has grown at less than half the rate projected in the early seventies. In the United States, electricity use grew at an annual rate of 7.5 percent between 1963 and 1973, and at an annual rate of 2.3 percent between 1973 and 1983. In France, electricity growth averaged 3.9 percent during the past decade, in Japan it averaged 2.5 percent, and in West Germany it averaged 2.4 percent. In Great Britain, electricity use in 1983 was slightly lower than in 1973, after a three-year decline between 1979 and 1982. Growth in the per capita use of electricity has slowed dramatically in most industrial countries in the past decade.[27] (See Table 1.)

There are several reasons for the less rapid increase in electricity use in industrial countries. A major cause is slower economic growth during the seventies. But something else is at work as well. In the

Table 1: Electricity Use Per Capita and Rate of Growth In Selected Countries

Year	France		West Germany		United Kingdom		United States	
	(kwh)	(percent)	(kwh)	(percent)	(kwh)	(percent)	(kwh)	(percent)
1962	1,598	—	2,180	—	2,577	—	4,187	—
1967	2,118	5.8	2,800	5.1	3,257	4.8	5,565	5.9
1972	2,838	6.0	4,095	7.9	4,044	4.4	7,621	6.5
1977	3,615	5.0	4,969	3.9	4,320	1.3	8,863	3.1
1982	4,480	4.4	5,424	1.8	4,173	−.7	9,011	3.0

Sources: United Nations and OECD.

United States, for example, electricity use since the mid-seventies has barely kept pace with growth of the gross national product. In earlier years electricity growth consistently exceeded economic growth by a substantial margin.[28] (See Figure 1.) New uses for electricity are not as large as they once were, and higher electricity prices are encouraging

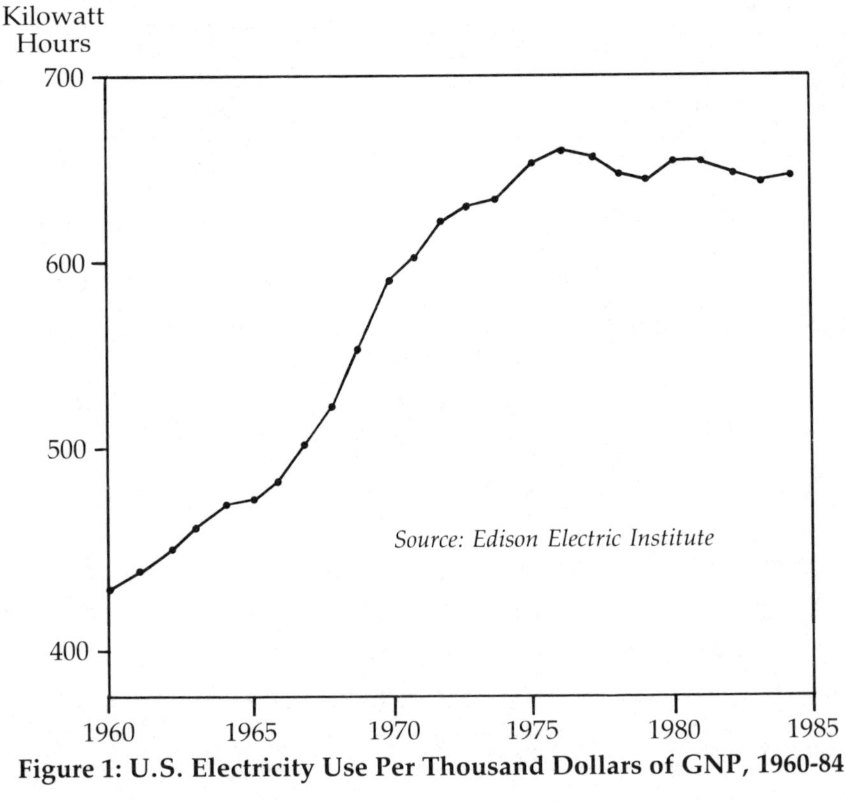

Figure 1: U.S. Electricity Use Per Thousand Dollars of GNP, 1960-84

conservation. Electricity's boom days may be gone for good, and if so, new approaches will be needed.

New Beginnings

With hundreds of billions of dollars of fixed investments and close ties to governments around the world, it is not surprising that the electricity business resists rapid change. But crisis often leads to new opportunities. Though the utility industry confronts unprecedented economic and environmental challenges, it is also being shaped by some of the most exciting innovations in its history. Nowhere is this more apparent than in the United States.

Irwin Stelzer, president of National Economic Research Associates, a leading consultant to U.S. utilities, says, "By the usual Wall Street indexes, the electric utility industry is a financial invalid . . . "[29] The problem is one of swollen construction budgets at a time of high interest rates and slower demand growth. Electricity is the world's most capital-intensive industry. Spending in the United States is expected to total $158.6 billion between 1983 and 1987, more than in the auto, chemicals, and petroleum industries combined. In most states these investments cannot be fully charged to customers until the plants are operating. As a result, the utility industry's long-term debt rose from $42 billion in 1972 to $125 billion in 1982. Annual interest charges alone reached $11.5 billion in 1982.[30]

Adding to utility woes are falling stock prices and bond ratings that make raising capital even more difficult. Once considered "blue chip" investments for "widows and orphans," many utility stocks have become big losers. As stock was issued to pay for burgeoning construction programs, its worth was diluted, and stock prices fell well below their book value. Investors in utility stocks and bonds have recently insisted on annual yields as high as 20 percent to compensate for the high risks. In some cases, where partly built nuclear plants have been scrapped, the investor is left with "junk bonds" that trade for a small fraction of their face value.[31]

The 40 or so U.S. nuclear plants still under construction have a total sunk cost of about $75 billion. If all are completed they will, by 1990, add at least $120 billion to the country's "rate base," or close to half the value of existing utility plants.[32] And though nuclear plants have caused the worst problems, cost overruns on coal plants have also damaged utility finances. Utilities with the most troublesome plants have cut dividends, laid off employees, and taken other cost-cutting measures to avert collapse. Many utilities that trimmed their construction programs are now in much better financial condition. Contrary to past patterns, slow growth now appears to be a key to utility success on Wall Street.

The magnitude of proposed rate increases has upset the traditional policy of automatically requiring consumers to pay for any plant a utility completes. Regulators are now asking if investments are prudent and if cost overruns are caused by mismanagement. Tens of billions of dollars of proposed rate increases will be at stake in utility hearings in the next few years, and an estimated 35 million American families, almost one-third of the population, could be affected. Industries are also being hurt by the rate increases. In Missouri, for example, two nearly complete nuclear plants will cost the state an estimated 40,000 jobs as companies are forced to move to areas where electricity is affordable.[33]

Regulators are now thinking the unthinkable—allowing utilities to go bankrupt rather than passing unjustified costs on to consumers. Utilities such as Public Service of Indiana, Consumers Power of Michigan, and Public Service of New Hampshire have almost completed plants that may never operate due to technical and financial problems. The value of these plants approaches or exceeds the worth of all of the companies' other assets, and could threaten company survival.[34]

Although bankruptcy is a solution of last resort, some government officials and consumer groups say it is not as drastic as it seems. Power plants would not shut down, and the companies would prob-

> **"Contrary to past patterns, slow growth now appears to be a key to utility success on Wall Street."**

ably be restructured, perhaps under state or municipal ownership. Governor Mario Cuomo of New York, for example, believes that the Long Island Lighting Company, builder of the Shoreham nuclear plant, should be permitted to go bankrupt rather than commit the state to a multi-billion dollar bailout. Cuomo and others say that only bankruptcy ensures that utility shareholders pay their rightful share of the costs. Company executives, on the other hand, argue that their problems are caused mainly by government regulation and general economic woes, and that consumers will be better off if they pay for the projects quickly and allow utilities to continue construction.[35]

Though higher construction costs and tighter financial constraints are nearly universal, no other country's electric industry has experienced a financial crisis approaching that in the United States. This is partly due to the poor planning and management of many U.S. projects. But also important are the wide institutional differences between countries. Throughout much of Europe, utilities are state-owned and can charge taxpayers or electricity consumers for all expenses. An interesting example is the French state utility, which has built up a debt of $19 billion in recent years due to its large nuclear construction program. The pace of ordering has slowed but Electricité de France, because it is backed by the national treasury, has been able to continue building nuclear plants, despite recent projections that the power is not needed.[36]

The common problem faced by electricity planners today is how to deal with uncertainty. Because of the five-to-ten-year lead times typically required to design and build a new power plant, planners must rely heavily on long-range forecasts. Throughout the sixties and early seventies, most planners used a technique called "trend line forecasting." Clark Gellings, a forecaster with the Electric Power Research Institute, says, "Back in the sixties you plotted a few points on a piece of paper and trended them out, using simple regression technique to fit a curve. There seemed to be no limit to growth then."[37] This approach worked well at the time, but it could not anticipate change and utterly failed to predict the more erratic developments of the seventies and early eighties. Forecasters took almost a decade to catch up with changes after they occurred. Increasingly sophisticated econometric models and large computers accounted for many variables, including demographics, GNP growth, electricity prices, and structural changes in the economy. But as with any model, results are only as good as the assumptions that lead to them. The econometric forecasts remained biased in favor of past trends and misread new developments.

A striking example is the annual forecasts of the North American Electric Reliability Council, used by utility planners throughout the United States and Canada. For ten consecutive years between 1973 and 1983, the Council lowered its forecast, invalidating the previous year's efforts almost before the ink had dried, and calling into question billions of dollars of investments that had been premised on the previous year's predictions.[38] In 1983, summer peak demand in the United States was 40 percent lower than projected a decade earlier. This equals the output from 300 large nuclear plants that would cost $750 billion at current prices.[39] (See Figure 2.)

A rule of thumb for electricity planners is that generating capacity should exceed peak demand by about 20 percent. But due to forecasting errors, most utilities in industrial countries now have "reserve margins" of 30 to 50 percent. In some cases a large share of the idle generating capacity is oil- or gas-fired, and because these plants have relatively low construction costs and high operating costs, the overall penalty is not too great. But in general, excess plant capacity is an expensive luxury that the industry cannot afford. Electricity prices rise as larger investments are spread over fewer kilowatt-hours of power use.

In France, where the nuclear construction program has far outstripped growth in power use, a dozen relatively new coal-fired plants are being decommissioned. In the United States, utilities have responded to changing conditions by canceling 53 nuclear plants and 49 coal plants since 1980. During the same period, only 20 plants were

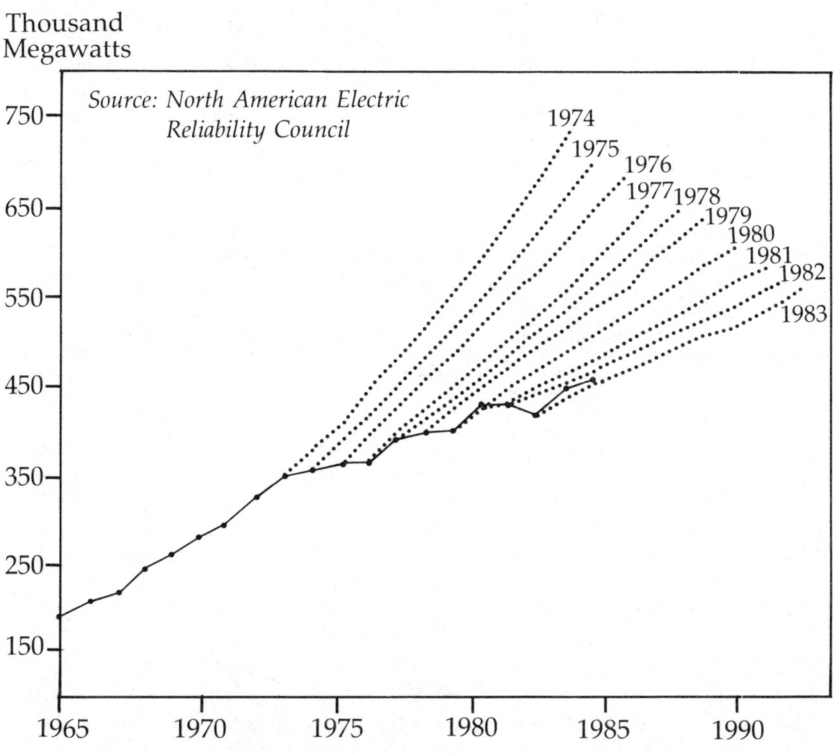

Figure 2: Summer Peak Electric Demand, 1965-84, And Projections Made from 1974 to 1983

ordered, all of them coal-fired. (See Table 2.) However, these cancellations were too little too late, leaving utilities with considerable excess capacity and massive bills for uncompleted projects. About $17 billion had been invested in planning and building the 53 U.S. nuclear plants canceled since 1980.[40]

Though electricity forecasting has grown more sophisticated, analysts still reach remarkably divergent conclusions, reflecting political and institutional biases and continuing uncertainty about which assumptions are valid. In the United States, the Edison Electric Institute and the Department of Energy predict that electricity use will grow at a rate of 3 to 4 percent annually during the next decade, roughly tracking projected economic growth.[41] Even this range represents a differ-

Table 2: Coal and Nuclear Plant Orders and Cancellations in the United States, 1970-84

	Orders				Cancellations			
Year	Coal		Nuclear		Coal		Nuclear	
	(plants)	(MW)	(plants)	(MW)	(plants)	(MW)	(plants)	(MW)
1970	25	12,442	14	14,275	0	0	0	0
1971	18	7,811	21	20,876	0	0	0	0
1972	27	12,682	38	41,526	0	0	6	5,738
1973	40	22,615	41	46,827	0	0	0	0
1974	71	34,183	26	30,931	0	0	8	8,290
1975	20	11,389	4	4,180	0	0	11	12,291
1976	13	5,938	3	3,790	2	800	2	2,328
1977	24	12,172	4	5,040	11	4,859	9	9,862
1978	28	14,634	2	2,240	5	3,125	13	13,333
1979	20	8,159	0	0	8	4,903	8	9,476
1980	6	2,688	0	0	9	4,348	16	18,085
1981	13	8,135	0	0	1	640	6	4,811
1982	1	600	0	0	0	0	18	22,019
1983	0	0	0	0	21	6,554	6	6,038
1984	1	572	0	0	18	7,923	6	6,780

Sources: Atomic Industrial Forum, Energy Information Administration, and Kidder, Peabody, and Co.

ence of 60,000 megawatts by the mid-nineties, equal to the output of 60 large nuclear plants. Other forecasters' predictions range from 6 percent annual growth to an actual decline in electricity use during the next decade.

Those who conclude that electricity use will grow rapidly usually make three key assumptions. They believe that the world economy will enjoy a period of sustained growth, that fossil fuel prices will generally rise faster than electricity prices, and that major new uses for electricity will emerge in industry, homes, and transportation. Harvard economist Peter Navarro writes that, "The bulk of energy-saving and productivity enhancing technologies—computers, telecommunications systems, and word-processing equipment—are electricity intensive." Energy consultants John Siegel and John Sillin similarly conclude that a sudden surge in power use will likely catch utilities off guard and result in power shortages by 1988.[42]

Projections of slower growth in electricity use hinge on assumptions that power will be more expensive in the future, that new uses such as microelectronics and electric steel mills will make only modest claims on power supplies, and that efficiency improvements will enhance electricity's productivity and moderate future growth. The most comprehensive U.S. study to date, a 1981 analysis by the Solar Energy Research Institute, concluded that technologies then available could allow a decline in U.S. electricity use through the year 2000, even with rapid economic growth. Similar conclusions have been reached by detailed studies in Great Britain, Sweden, and West Germany.[43]

Energy analyst Amory Lovins, who has argued for nearly a decade that heavy use of electricity makes little economic sense, stated in 1984 Congressional testimony that, "The critical question is not whether the potential for such vast improvements in America's electrical productivity exist—that seems beyond dispute—but how fast that opportunity will be seized."[44] This is indeed a critical question. Recent improvements in the efficiency of electric motors, light bulbs, and many other technologies indicate that the potential for reducing electricity use is enormous. Detailed end-use based forecasts now used by utilities in California and the Pacific Northwest generally show slower rates of electricity growth. But since many consumers lack information, and marketing new technologies takes time, realizing this potential savings is not guaranteed.

Although debates about electricity growth often seem arcane, they are more than academic. The U.S. Department of Energy spent $2.5 million on a 1983 study warning that the country risked running short of electricity unless a major program of power plant construction began immediately—at a cost of about $1 trillion (1982 dollars) by the year 2000.[45] Throughout the world, government leaders and utility executives are considering building hundreds of new coal and nuclear plants. In addition to their cost, these plants would set the course of many nations' energy futures for decades to come and could risk major harm to the global environment.

Utility industry leaders frequently paint an apocalyptic vision of a future without hundreds more coal and nuclear plants. Harold Finger, President of the U.S. Committee for Energy Awareness, said in 1984 that, "Without enough electric power to support a strong base of conventional industry, we will in effect undermine our national goals of international competition, economic growth and world leadership." Eugene Oatman of the Electric Power Research Institute gave a 1984 speech entitled, "If the Lights Go Out? — The Day After."[46]

These arguments miss a key message of the diverging electricity forecasts. Electricity's role in the next decade or two is more uncertain than at any time in the recent past. Utility strategies should be geared to reducing uncertainty, minimizing costs, and ensuring that adequate power is available regardless of whose econometric model turns out to be more finely tuned. Building a large coal or nuclear plant requires 6 to 15 years and costs $1 billion to $5 billion. Not only is this beyond the time range over which analysts can make accurate

> **"Electricity's role in the next decade or two is more uncertain than at any time in the recent past."**

forecasts, but the very uncertainty as to the cost and length of the projects further confounds planners.

Today smaller plants and incremental strategies have an inherent advantage over massive construction projects. A 50-megawatt project, for example, might take one-third the time to build that a 1,000-megawatt project does, and if demand continues to rise in the interim, additional small units can be added. Similarly, houses can be insulated and industrial motors replaced by more efficient ones in a matter of months. Financing for small projects that can be built rapidly is less burdensome. Even at a somewhat higher cost, small-scale power projects and efficiency investments may deserve priority because of the flexibility they provide.

The potential now exists for major advances in the electric industry. They are by no means guaranteed, however, since most countries still favor large power projects through a variety of tax subsidies, special loans, strict monopolies on power generation, and disincentives for efficiency investments. Ignoring the potential would risk a future of high-cost electricity, financially hemorrhaging utility companies and damage to the global environment and human health.

Small-Scale Power Production

Of all the energy laws passed during the seventies, the U.S. Public Utility Regulatory Policies Act (PURPA) may have the most far-reaching consequences. Part of the PURPA law, which caused little controversy when passed in 1978, could end the traditional utility monopoly over power generation. PURPA directs utilities to interconnect with small-scale independent power producers (also known as "qualifying facilities") and to pay a fair market price for the electricity. More than any government research and development program, tax subsidy or loan guarantee, PURPA has brought a new spirit of entrepreneurialism to the utility industry. Prompted by PURPA, several hundred U.S. companies have entered the power generation business since 1980, working alongside established companies to harness new technologies.[47]

Until recently the utility industry appeared mature, with little progress expected. But apparent stagnation in power generation technology is probably due more to neglect of research and development than to any inherent characteristics of the technologies. Materials science, semiconductor physics, and even aerodynamics and biotechnology are now leading to major efficiency improvements and cost reductions in power generation. Most of the advances have been made not in conventional thermal power plants but in a range of heretofore neglected technologies.

Cogeneration—the combined production of heat and power—is the predominant new electricity source being developed. Although industrial cogeneration supplied half of U.S. electricity at the turn of the century, by the seventies its share fell to a mere 3 percent. Cogeneration is somewhat more common in Europe, thanks partly to the prevalence of district heating plants that employ the technology. West Germany and Finland each get about one-quarter of their electricity from cogeneration; France and Italy about 18 percent. Revived interest in cogeneration has centered mainly in the United States, where it now provides over 15,000 megawatts of power. Another 200 projects with a total generating capacity of 6,000 megawatts are under construction.[48]

Two kinds of energy are needed in most industries: electricity and heat (usually in the form of steam). In recent decades most companies have produced their own steam (using an oil- or natural gas–fired boiler) and purchased electricity from a local utility. Steam production is 90 percent efficient, but electricity generation and transmission only capture one-third of a fuel's energy value, making electricity a more expensive form of energy. The overall energy efficiency of a typical industrial plant producing steam and purchasing electricity is usually between 50 and 70 percent.[49]

With cogeneration an industry can often raise the total efficiency of its plant to between 80 and 90 percent, reducing energy costs substantially. In most systems, the low-pressure boiler used to generate process steam is replaced by a high-pressure boiler that powers a steam turbine and electric generator. The low-pressure steam exhausted from the turbine is used for industrial heat, space heating and cooling, and water heating. Electricity generation using this approach consumes only half as much fuel as a conventional power plant.

Only when large and relatively constant amounts of heat are required is cogeneration economical. If heat is needed only for a few hours a day or just during winter, a cogeneration system would either stand idle much of the time or operate inefficiently, producing waste heat the way a conventional power plant does. But many factories operate 16 to 24 hours daily, making them promising candidates for cogeneration. The pulp and paper, chemical, primary metals, refining, and food processing industries all have large heat requirements and have recently begun installing cogeneration systems. A large market for cogeneration also exists in aging oil fields, where heat is needed to help recover remaining reserves.

Since cogeneration is based on existing technologies already worth billions of dollars, industry has expanded research and development and commercialized the process relatively quickly. Major corporations such as Westinghouse and General Electric have been joined by dozens of smaller companies such as Thermoelectron and Applied Energy Systems in attempting to develop this market. Often, cogeneration projects are joint ventures undertaken by the host company, an outside firm that manages the project, and occasionally a utility company. Sometimes the facility is actually owned by an outside developer that sells steam to the host company and electricity to the local utility. To succeed these complicated projects require cooperation among financiers, lawyers, and contracting companies.

Cogeneration technology has advanced rapidly since the mid-seventies. Frank DiNoto of Hawker Siddely Power Engineering notes,

"Right now its the only business of any consequence in power equipment." Today's cogeneration systems are more efficient and run on a wider variety of fuels. Many cogenerators employ efficient diesel engines and gas turbines that are driven by exhaust gases rather than steam, allowing higher efficiency. Unfortunately, these systems must be fueled with natural gas or petroleum, both expensive, premium fuels. Research on burning coal, wood, and agricultural wastes is under way and shows considerable promise. Small fluidized-bed coal plants that emit less pollution may well be commercialized within a few years. Several cogeneration plants fueled by wood or urban wastes have already been built.[50]

One of the largest cogeneration projects to date provides power for the Dow Chemical Company plant in Freeport, Texas. This huge facility has relied partly on cogeneration since the forties. Rising fuel and electricity prices in the late seventies made it economical to replace much of the company's antiquated cogeneration equipment and install additional capacity. Dow Chemical now has 1,300 megawatts of cogeneration. It uses the heat to process chemicals and sells some of the electricity to Houston Lighting and Power. The large petrochemical industry in Houston has thousands of megawatts of cogeneration potential, and several companies are rapidly developing it. Diamond Shamrock has signed a $1.3 billion, ten-year contract with the utility to sell 225 megawatts of cogenerated power. Big Three Industries is at work on a 300-megawatt cogeneration plant that will sell steam to several Houston-based companies.[51]

The energy-intensive pulp and paper industry is another cogeneration leader. This industry's traditional heavy reliance on its own waste products for fuel has been reinforced by rising oil and gas prices, and many of the new projects use cogeneration to boost energy efficiency. The Scott Paper Company, for example, has installed cogeneration equipment at one-third of its paper mills, stretching from Maine to Alabama. In Mobile, Alabama, the company is building turbine generators and biomass and waste heat recovery boilers to make itself 60 percent energy self-sufficient. Excess power will be sold

"The Scott Paper Company has installed cogeneration equipment at one-third of its paper mills."

to the Alabama Power Company. This $300 million project is the largest single capital investment in Scott Paper's history.[52]

Most cogeneration projects begun so far range from 10 megawatts to 300 megawatts, but much smaller systems may soon be economical. Large cogeneration plants are usually custom-designed and much of the equipment is built on site. Building a small cogeneration plant the same way would be prohibitively expensive. Only mass production of modular systems will make small-scale cogeneration affordable. Several companies are building such packaged systems, but they have not been widely marketed. About 40 systems were sold in 1983 and some 200 in 1984, according to the Frost and Sullivan market research company.[53]

A particularly promising system is a 65-kilowatt diesel cogeneration plant fueled by natural gas and designed by Hawthorne Energy Systems of California for the McDonald's restaurant company. It produces electricity as well as heat to run a restaurant's hot water and air-conditioning systems. A specially-designed microelectronic chip programmed with climatic and economic data continuously adjusts the system in response to the weather, energy requirements, and the utility's price for cogenerated power. Although McDonald's has only installed one of these systems, its performance so far has been excellent, and the company is considering ordering many more. Engineers believe that similar systems installed in quantity would have a payback period of four years or less in areas where electricity prices are high. If installed in other fast food restaurants, as well as grocery stores, shopping malls, hospitals, and schools, small-scale cogeneration systems would find a market worth billions of dollars and add significantly to energy supplies.[54]

Cogeneration systems are more economical than virtually any other power source available. Installed costs range from $500 to $1,000 per kilowatt, depending on the technology and fuel.[55] Total generating costs are less than half those for nuclear plants being built and one-fifth less than coal costs. Surveys estimate that the United States could someday harness between 100,000 and 200,000 megawatts of cogenerated power, or as much as one-third of current generating capacity.[56] But developing this potential will take time since cogeneration technologies are still evolving and many utilities discourage their use.

About 10,000 megawatts of cogeneration are planned in the United States, and total installed capacity should reach 20,000 to 25,000 megawatts by 1990. Between 25,000 and 50,000 megawatts are projected for the year 2000.[57] Although these forecasts seem ambitious, rapid advances in cogeneration in the past two years may prove them to be conservative. Frost and Sullivan projects that as many as 40,000 small-scale, modular cogeneration systems will be operating in the United States by the year 2000.[58]

Wind energy appears likely to join cogeneration as a major new power source. Starting from near zero in 1981, about 9,000 wind machines with a generating capacity of over 700 megawatts have been installed in California in the past three years.[59] (See Table 3.) Virtually all are installed at wind farms, clusters of machines located in mountain passes and connected to utility lines. These wind machines churned out enough electricity in 1984 to supply 70,000 homes, marking the first time that wind energy has made a significant contribution to a modern utility grid.

Substantial research and development by a dozen small companies, largely without direct government support, has led to the many reliable and economical wind machines now being produced. Incorporating microelectronic controls, aerospace concepts, and a host of modern materials and engineering principles, these new wind machines are a major improvement over older wind power technologies. Already wind machines are being routinely installed at a cost of $1,500 to $2,000 per kilowatt. Generating costs are estimated at between 10-15¢ per kilowatt-hour. But modern wind power technology is still unfolding, and costs should fall to less than $1,000 per kilowatt, or 6-8¢ per kilowatt-hour in the next few years.[60] The California

"Modern wind power technology is still evolving and costs should fall to less than $1,000 per kilowatt."

Table 3: California Wind Farms, 1981-84

	Machines Installed	Capacity Installed	Average Capacity	Average Cost	Power Generated[1]
		(megawatts)	(kilowatts)	(dollars/ kilowatt)	(million kilowatt hours)
1981	144	7	49	3,100	< 1
1982	1,289	64	50	2,175	6
1983	2,816	189	67	1,900	74
1984[2]	4,990	480	96	1,600	700
Total	9,240	740	80	—	780

[1]Most wind machines are installed in the last half of a given year and do not produce substantial power until the next year. [2]Preliminary estimate.

Source: California Energy Commission.

Energy Commission projects that wind power will be the state's second least expensive power source by 1990—right behind hydropower.

More innovative than the technology are the business arrangements designed to harness wind power. Wind farm developers purchase or lease land in windy areas, manufacture or buy wind machines, raise capital from investors who can take advantage of state and federal tax credits, and sign a standard contract with the local utility to sell it power for 10 to 20 years. The California government requires utilities to establish regular procedures and fair prices for interconnecting with small-scale power producers. Tax credits have been essential to the economic viability of wind farms so far, but will not be needed within a few years.

The more reliable and less expensive wind machines being built in California will lay the groundwork for wind power in parts of the world with less ideal wind conditions. If other states and countries provide similar opportunities for private companies to enter the wind farm business and become independent power producers, wind energy could supply 10 percent or more of the power in many areas by the end of the century. Some limited wind farming has begun in the states of Hawaii, Montana, New York, and Oregon, as well as the New England states. Wind farms are also being planned in Denmark, Great Britain, the Netherlands, Sweden, and several islands in the Caribbean.[61]

Geothermal energy is another new source of electricity, though its potential is limited by the relatively small number of high-quality reserves of subsurface steam and hot water. Where high-pressure steam is near the surface, geothermal power generation is already a bargain. At the Geysers in northern California, over 20 separate power plants have been installed in the past decade and together provide 1,300 megawatts of power. A geothermal plant uses a steam collection system, turbine, generator, and pollution control equipment, all relatively standard technology. Generating costs are reported as low as 5¢ per kilowatt-hour. The Philippines has developed four geothermal fields and is working on several more, hoping to have 1,700 megawatts of capacity by the end of 1985. Mexico has developed three major geothermal fields and now has a capacity of 645 megawatts.[62]

Central America, parts of Southeast Asia, and the western United States have the potential for major reliance on geothermal energy. Prime sites also exist in parts of southern Europe and East Africa. An international survey by Ronald DiPippo of Southeastern Massachusetts University estimates that 10,000 megawatts-worth of geothermal power plants will be in place by 1990. This will require new technologies to tap deeper geothermal reservoirs and use lower temperature geothermal water. If these technologies are developed, geothermal energy use could reach 30,000 to 50,000 megawatts by the end of the century.[63]

Small-scale hydroelectric generators, once major electricity sources, have fallen into disrepair in many countries in recent decades. In the past few years renovation schemes and newly built facilities have increased small-scale hydropower supplies in the United States by almost 300 megawatts. An equal amount of capacity is currently being built. The cost of a new facility ranges from $2,000 to $3,000 per kilowatt, but retrofitting an old dam is considerably less expensive. Rapid development of small-scale hydropower is occuring, but will likely be slowed by environmental constraints in many areas. Growth of this power source is likely to be most rapid in the Third World, where power is lacking in many rural areas. China has long relied on small-scale hydro to power rural communes and is now exporting its technology to other countries.[64]

More abundant are a host of biological fuels such as wood and agricultural wastes. Scores of small-scale biomass- and waste-fired power plants are now being built, mainly in North America and Scandinavia. The United States currently has about 1,400 megawatts of such capacity and another 1,500 megawatts planned or under construction. Close to half the total comes from wood wastes, mainly burned at wood industry plants that generate their own power and sell the excess to a utility company. Plants fueled by agricultural and municipal waste are rapidly growing in importance. About 90 plants are currently planned or being built. Garbage can either be burned directly or methane can be extracted from a landfill and used to run a generator.[65]

In Burlington, Vermont, the municipal utility completed a 50-megawatt wood-fired power plant in 1984. It is now the world's largest and cheaper than an equivalent coal-fired plant. Other projects are being designed and financed by wood products companies, many of them using cogeneration to harness heat as well as electricity. Small independent companies similar to those developing wind farms are leading the way in biomass projects. One such firm is the Ultrasystems Company, which is building several 10-megawatt-plus biomass-fueled power plants. Ultrasystems oversees the construction and operation of the plant and raises the capital needed to complete it. The power is sold to a utility company under a long-term contract. One such project, a 24-megawatt, $40 million plant to be fueled by forest residues, was begun in central California in 1984.[66]

Wood-fired power plants are currently being built for less than $2,000 per kilowatt. In areas with abundant wood supplies these plants have generating costs of under 7¢ per kilowatt-hour, competitive with most alternatives available. Wood- and waste-fired power present few technical or economic obstacles. Much of the challenge comes in ferreting out the abundant but dispersed waste products that can serve as feedstock. The biomass power industry plays a useful role in locating these materials and putting together the technology and financing needed to harness a new power source. A New York dairy farmer who has signed a contract with a company that will build a power plant on his farm says, "It took me 40 years to learn how to make cottage cheese. I don't want to start learning how to make electricity." The mounting problem of disposing of urban wastes has led many municipal governments to welcome such projects and even pay a substantial fee to a company willing to remove the wastes.[67]

No good estimates are available of the potential for using biomass in electricity generation, but waste products ranging from forest residues to walnut shells are abundant everywhere. In the United States, development so far has been concentrated in the Southeast, the West Coast, and New England. Sweden leads in harnessing wood-fired energy, mostly for district heating plants that use cogeneration. The Philippines has built 17 wood-fired power plants since the late seventies and plans to have 60 by 1990.[68] Each has a capacity of 3.3 megawatts and is fueled by a plantation of fast-growing trees. Together the plants will be a substantial component of the country's power system in the nineties.

Other energy technologies not yet ready for major commercial use may have even greater long-run potential. Solar electricity produced by photovoltaic cells is one promising power source. Solar cells can be

"The Philippines has built 17 wood-fired power plants since the late seventies."

installed at generating plants in rural areas or on rooftops, and will allow a much greater decentralization of electricity supplies than virtually any other technology. Costs must fall to about one-fifth their current level to be competitive with utility power, but projections indicate that this may occur by the nineties if research funding is kept high. In addition, solar thermal power technologies and solar ponds are projected to have competitive generating costs, at least in sunny climates, within a decade. Fuel cells that run on natural gas, hydrogen, or some other fuel are now projected to be a practical household or industry energy technology. Installed in a basement, they could heat and cool a home as well as produce electricity.[69]

Small-scale power production using a variety of new energy sources is taking hold far more rapidly than projected a few years ago. Half of all U.S. utilities now obtain some of their power from independent energy producers. Figures from the U.S. Federal Energy Regulatory Commission show that since 1980 applications have been filed for 785 small-scale power projects with a total generating capacity of 14,193 megawatts. (See Table 4.) The average power output of each plant is an extraordinarily low 18 megawatts, less than 2 percent of the capacity of a standard nuclear plant. Cogeneration is the most important component, accounting for two-thirds of the total, but each year the mix of new sources becomes more diverse. Wood, small-scale hydroelectric, and wind projects are growing most rapidly.

If small-scale power projects continue to be launched at the pace of the past two years, the United States alone would obtain 60,000 megawatts from them by the end of the century, or about as much as nuclear power now provides. Other countries that have not yet pursued small-scale power generation are likely to have similar potential. In developing countries, where populations are more dispersed and electric grids do not reach many areas, some of these technologies are likely to be particularly appropriate. Conditions are ripe for a rapid increase in reliance on small-scale power sources if the institutional hurdles and biases are cleared away.

Table 4: Independent Power Projects Planned in the United States, 1980-84[1]

Source	1980	1981	1982	1983	1984	Total
			(megawatts)			
Cogeneration	319	844	2,818	3,211	2,531	9,723
Biomass[2]	0	235	534	401	616	1,786
Hydro	59	45	63	380	382	929
Wind	76	24	32	340	384	856
Geothermal	76	80	76	65	203	500
Waste	1	0	0	124	171	296
Solar	0	0	0	87	16	103
Total	531	1,228	3,523	4,608	4,303	14,193

[1]Includes projects for which applications have been filed with The Federal Energy Regulatory Commission. [2]Includes wood and agricultural wastes.

Sources: *Cogeneration & Small Power Monthly;* Worldwatch Institute

Energy Efficiency as a Power Source

Utilities have long regarded improved energy efficiency as an unwanted competitor that cuts into electricity sales. Amid aggressive campaigns to promote air-conditioning and all-electric homes, utilities and other companies have neglected research on how to reduce the power requirements of electrical motors, household appliances, and dozens of other technologies. But today much more efficient technologies have been developed and make far better investments than do new power plants.

About one-third of the electricity generated in industrial countries powers household appliances. As electricity prices have increased, the average efficiency of new appliances has also begun to rise—by more than 50 percent in Japan, but by only 10 to 20 percent in the United States, which started with somewhat more efficient appliances. Potential efficiency is far higher. A 1983 study by Howard Geller of the American Council for an Energy-Efficient Economy found that the most efficient refrigerator then available used one-quarter less power than the average model sold, the most efficient central air-conditioner used 40 percent less power, and the most efficent electric water heater used two-thirds less power.[70]

If all U.S. appliances were replaced by the most efficient models, summer peak electricity demand would fall by about 75,000 megawatts, more than current nuclear capacity and equal to eight years of demand growth. The extra cost of these efficient appliances compared to the average model is just 2-6¢ per kilowatt-hour saved, or less than the electricity cost from virtually any power plant now being built.[71] Because companies bring more efficient models to the marketplace each year, future savings would be even greater.

Lighting is another major electricity consumer, accounting for more than 20 percent of the total in many countries. Light bulb manufacturers have been working for almost a decade to improve bulb efficiency, with some success. Incandescent light bulbs similar to those used in most homes, but requiring 10 to 15 percent less power, are now available. But incandescent lights are inherently inefficient, turning about 90 percent of the electricity they use into worthless heat. Incandescent bulbs use 40 percent of U.S. lighting energy but supply only 16 percent of the light.[72]

Fluorescent light bulbs are more than three times as efficient, but they produce a flat white light and require a special lighting ballast to regulate the current they receive. As a result, fluorescent lighting is confined mainly to commercial buildings. Recently, however, engineers have designed fluorescent bulbs that plug into an ordinary socket and produce a more pleasing light. Metal halide lights have been developed that are even more efficient, and are now being introduced in commercial buildings. Improved bulbs, along with more effective use of natural light and electronic control of lighting levels, could probably cut the electricity used for lighting by more than half, reducing national electricity use 10 percent. But few people consider the energy requirements of lightbulbs when they buy them, and the limited use of more efficient lightbulbs in recent years indicates that some form of government standards or incentives are needed.[73]

Insulation, storm windows, and other conservation measures have an enormous potential to reduce electricity use in buildings. In the United States half of all new houses are electrically heated, and power use for space heating is expected to rise 60 percent by the end of the century. The heating efficiency of homes in Western Europe and the United States has improved 20 to 40 percent over the past decade, but is still far from its potential. Swedish homes already use 30 to 50 percent less heat than American homes of the same size, and some contractors in Canada and Sweden routinely build homes that require little if any supplementary heating or cooling, even in the harshest climates. Such homes cost less than 5 percent more than conventional homes and pay back their efficiency investment in two to three years. Similar improvements are possible in large apartment and commercial buildings that use electronic energy management systems to optimize heating, cooling, and lighting.[74]

Industry accounts for close to half of worldwide electricity use and for a much larger share in developing countries. Power requirements are particularly high in large materials processing industries such as cement, chemicals, and metals. And while use of oil and natural gas in industry has fallen, the use of electricity has grown at more than 2 percent per year in the past decade. Substituting electricity for fuels can in some cases greatly boost end-use efficiency. Electric arc steel mills that process scrap steel are rapidly replacing traditional mills powered by metallurgical coal. Aluminum smelting, which uses 45,000 megawatts of electricity worldwide, is growing, but technologies can cut electricity requirements for aluminum production by

"Installing more efficient technologies would require less than half the investment needed to get an equivalent amount of power from new plants."

25 percent. Recycling aluminum saves a full 90 percent of the power needed to produce it.[75]

Almost two-thirds of the power used in industry runs electric motors, and until recently little had been done to improve their efficiency. Now motors with improved designs are being introduced in a wide variety of sizes. Much greater savings will be gained with electronic adjustable speed drives that reduce electricity use in motors 30 to 50 percent. Only about 100,000 such motors now operate in the United States, but their use is growing rapidly. Overall, efficient electric motors could probably reduce power use in most countries by at least 10 percent.[76]

For some applications, electricity is simply an inappropriate energy source. For example, space heating with electric resistance heaters is extremely inefficient, a practice Amory Lovins has likened to cutting butter with a chain saw. Efficient gas-fired furnaces or electric- or gas-fired heat pumps would be both more economical and fuel-efficient. Yet the electric industry actively promotes some of the least sensible uses of electricity. Electricité de France, for example, promotes electric space heating as a way of absorbing the large oversupply of nuclear power the country is now committed to. But once electric heating is installed in a home, the owner may well be stuck with decades-worth of rising energy bills, since replacing a home's heating system is extremely expensive.

Opportunities for using electricity more productively are seemingly limitless. In most industrial countries, even with economic growth at a healthy 4 percent annual rate, electricity use need not exceed the current level.[77] Installing more efficient technologies would require less than half the investment needed to get an equivalent amount of power from new plants. Opportunities for saving electricity are equally great in developing countries, though rapidly expanding cities and industries will in most cases still require new power supplies.

An established relationship with millions of electricity consumers and ready access to capital markets puts utilities in a pivotal role in promoting conservation. When utilities are actively involved, expensive construction programs can often be avoided. Utilities can effectively encourage conservation by supplying information, offering rates that reflect the real cost of generating power, and providing financial assistance. But it is important that utilities not develop an uncompetitive monopoly in conservation. Much of the recent progress made in energy efficiency has come through the pioneering efforts of energy service companies that reduce the energy bills of an office building or industrial plant for an agreed upon price. Utility conservation programs should assist rather than supplant these efforts.

Since the mid-seventies many U.S. utilities have adopted conservation programs, mostly in response to government pressure. The federal Residential Conservation Service created by Congress in the late seventies requires utilities to offer energy audits to residential customers, and many states mandate much more substantial conservation efforts. A growing number of state utility commissions allow utilities that make conservation investments to include these sums in the "rate base," just as they would an investment in a new power plant.

A 1982 survey by the Investor Responsibility Research Center found that 72 percent of U.S. utilities have formal energy conservation programs while two-thirds have load management programs that redirect power use to off-peak hours. (See Table 5.) Noting that half of these have been established since 1980, the survey describes a "virtual stampede" by utilities to make conservation "a vital part of their overall operations."[78] The 120 utilities surveyed expect that their peak load can be reduced by 30,000 megawatts during the next decade, saving $19 billion in avoided construction at a cost of only $6 billion. Conservation programs include energy audits, home weatherization loans, and cash rebates for the purchase of energy-efficient appliances. A survey by Lawrence Berkeley Laboratory estimates that half

"Conservation programs include energy audits, home weatherization loans, and cash rebates for the purchase of energy-efficient appliances."

of all U.S. electricity consumers are served by a utility that offers a rebate on the purchase of energy-efficient appliances.[79]

Table 5: Largest U.S. Utility Efficiency and Load Management Programs

Company	1982 Generating Capacity	Planned Savings By 1992	Savings as Percent of 1982 Capacity	Projected Increase in Demand Through 1992[1]
	(megawatts)		(percent)	(percent per year)
TVA	32,076	4,000	12	2.4
Duke Power	14,526	2,994	21	3.9
Florida P&L	12,865	2,100	16	3.5
Pacific G&E	16,319	1,871	22	0.9
Pacific P&L	8,805	1,750	22	3.0
Houston L&P	12,966	1,700	13	2.6
So. Calif Edison	15,345	1,500	10	2.0
Florida Power	5,899	1,500	25	1.0
Public Srv. E&G	9,023	956	11	1.3
Bonneville Power	0	802	NA	NA
Jersey Central	3,371	800	24	1.5
Alabama Power	9,194	800	9	2.6
Penn Electric	2,736	671	25	2.0
Los Angeles DWP	6,749	601	9	1.7
Oklahoma G&E	5,359	600	11	NA

[1]Projections include the results of planned efficiency programs.

Source: *Generating Energy Alternatives*.

One of the more comprehensive energy management efforts, mounted by the Florida Power and Light Company, has a goal of reducing the area's peak power demand by 2,100 megawatts between 1982 and 1992—16 percent of the company's current generating capacity. So far the company has performed energy audits on almost 300,000 homes, encouraged the replacement of 50,000 inefficient central air conditioners and heating systems, upgraded the ceiling insulation of 31,000 homes, and tightened windows on 32,000 homes. Florida Power and Light provides cash rebates for replacement of inefficient electric elements in home water heaters and for businesses to purchase more efficient fluorescent light bulbs. The company also offers financial incentives to area merchants who sell energy-efficient appliances and reports that this has prompted the sale of 134,000 such appliances since 1982.[80]

Connecticut-based Northeast Utilities announced a plan in 1981 to avoid further construction of central power plants by adopting conservation and load management programs. The utility weatherizes customers' homes, charging for materials but not labor. It also offers subsidized loans for some conservation measures and provides energy audits. Carolina Power and Light has several conservation and load management programs, including special low-interest loans designed to shave power requirements by 1,750 megawatts over the next decade. In northern California the Pacific Gas and Electric Company has provided $168 million-worth of zero interest loans for customers who install specific conservation measures. It also performs 100,000 energy audits per year and provides rebates for some energy efficient equipment. The utility expects to spend $1 billion on these programs over the next decade, while reducing peak power use by 1,900 megawatts.[81]

Load management programs are not yet as extensive as conservation programs, but a growing number of utilities are adopting them. Most consist of lower prices for customers who use power during off-peak periods such as night or early morning hours, when power use is generally lower. This reduces a utility's peak demand, which is usually met with the least efficient and most expensive power sources, often oil- or gas-fired units. An extra kilowatt-hour at the peak can easily cost twice as much as the average kilowatt-hour generated at other periods. Remote controls capable of limiting the on-peak use of air-conditioners and other commercial and industrial equipment are increasingly being tried by utilities. Customers receive a special incentive electricity rate if they join such a program. Falling prices for microelectronic equipment may soon make it possible to regulate power use in most homes and apartments by remote control.

New England Electric has an extensive load management program. It has installed controls on most of its customers' electric water heaters, allowing them to be turned down during peak winter periods. "Time of use" rates are being adopted for most commercial and industrial customers. And a test program is under way asking a limited number of residential customers to turn down their heat during the most severe peak periods. The utility has found that during about 60 peak hours each year, it is economical to pay customers large sums for every kilowatt-hour not used. New England Electric is monitoring load growth carefully and plans to expand its load management program if it proves cost-effective.[82]

Though utility efficiency programs are growing and some have achieved impressive results, such efforts still have a long way to go. Many utilities do little more than make programs available; no real effort is made to encourage participation. Energy audits are often superficial, revealing only a small portion of the conservation potential. In some cases "success" is measured by changes in customer attitudes, rather than by how much electricity is saved. Rental units and apartment buildings, particularly those housing low-income people, have been left out of many programs. And many include only the simplest and cheapest measures, such as shower flow restrictors or a few extra inches of attic insulation. They ignore more substantial improvements such as triple-glazed windows or installation of a more efficient furnace, investments that would more than pay for themselves over the lives of most buildings.[83]

Efficiency programs will have to be greatly stepped up if they are to take their logical place as cost-effective alternatives to power plant construction. For this to happen, utility executives will have to realize that conservation programs are in their own self-interest. Ralph Mitchell, a former utility executive now working for an energy services company partly owned by a utility, says, "A compelling reason for entering this [conservation] business is that energy conservation opportunities will ultimately be captured with or without us." The efficiency potential is enormous and involves billions of dollars of future business, but even if utilities do not realize the potential, regulators may provide incentives that force them toward conservation. Michael Foley, chief economist with the National Association of Regulatory Utility Commissioners, says, "The general consensus of the regulatory community is that building power plants should be the last option."[84]

Electricity's Future

In 1983 the U.S. Department of Energy completed a study on the future of electricity in the United States. It concluded that the country would need an additional 438,000 megawatts of generating capacity by the end of the century, about two-thirds of current capacity. The report calls for a $1 trillion nuclear and coal construction program as the only way of preventing a power crisis. Ralph Cavanagh of the Natural Resources Defense Council terms the study a "blueprint for fiscal suicide." Yet the report accurately reflects the philosophy of many utility planners. Wedded to the challenges and choices of the past, they see the main problem as how to finance all the plants that are needed. The biggest obstacles are regulators and consumers who are squeamish about letting electric rates rise rapidly enough to pay for the plant construction. Missing from the study is any serious consideration of efficiency or small-scale power sources as alternatives to construction programs.[85]

Not all utility planners are mired in the past. The Southern California Edison Company stunned the utility world in 1980 by announcing

"Small-scale power sources, including cogeneration, wind power, and geothermal energy, will supply virtually all the new generating capacity in California."

plans to rely largely on new generating technologies and improved efficiency to meet future growth in electricity use. Four years later, energy efficiency investments have grown and hundreds of small-scale power projects will soon feed electricity into the company's grid. Most of the new power projects are owned not by Southern California Edison but rather by a new breed of independent energy producers.

New technologies for power generation and improved efficiency can no longer be dismissed as impractical or uneconomical. But still at issue is how rapidly the new energy sources will be developed and who will control them. In most parts of the world, institutional and financial obstacles continue to slow progress toward more decentralized and efficient electricity systems. Policymakers around the world need to redefine the role of utilities and how governments regulate them.

California is where changes in the electricity industry have first gained attention. By mid-1984 the state had over 10,000 megawatts of new small-scale generating sources planned or under construction, most of it by independent energy producers who will sell the power to the state's utilities. In Pacific Gas and Electric's service area in northern California, generating capacity equal to one-third of peak power use is in various stages of development. Southern California Edison is proceeding apace. (See Table 6.) Because both utilities are short of investment capital, they welcome the prospect of gaining new generating capacity without making major investments.

What began as little more than a token effort to encourage new energy sources has become the centerpiece of California's energy future. Aside from the long-complete but not yet operating Diablo Canyon nuclear plants, small-scale power sources, including cogeneration, wind power, and geothermal energy, will supply virtually all the new generating capacity in the state. No new coal or nuclear plants are planned. Although utility executives say they may still need central power plants in the nineties, that appears less likely with each passing

Table 6: Small-Scale Power and Cogeneration Technologies Planned by Selected Utilities in 1984[1]

	1983 Peak Load	Small-Scale Power Sources: Operating	Under Construction	Under Negotiation	Small-Scale Power Sources as Proportion of 1983 Peak Load
		(megawatts)			(percent)
Pacific Gas and Electric	15,156	684	2,198	2,155	33
Southern California Edison	13,464	552	1,718	1,848	31
Houston Lighting & Power	10,676	1,945	695	—	25

[1]Figures for mid-1984.
Source: Worldwatch Institute

month. New energy sources are now being developed in California at a pace that will not only meet projected growth in electricity demand but allow much of the state's fossil fuel–fired power generation to be phased out. Jan Hamrin, president of Independent Energy Producers, believes that her industry will supply 20 percent of the state's electricity within five years.[86]

California is far ahead in developing new generating sources for a number of reasons. Some renewable energy sources, such as wind and geothermal heat, are abundant. In addition, oil- and natural gas–fired plants account for half of California's generating capacity, and these expensive plants can be quickly and economically phased out as other power sources are developed. But more important, California's state government has actively promoted new energy sources. Policymakers in the state capital of Sacramento began rewriting the rules governing the state's utility industry even before PURPA had been pieced together in Washington, D.C.

The California Public Utilities Commission requires that utilities offer standard contracts to independent producers who want to sell power, a provision urged by Independent Energy Producers, a group representing the emerging small-scale power industry in California. Some contracts are short-term agreements based on a regular reading of natural gas–based generating costs. Others are long-term agreements based on what power would cost from plants built in the future. Interconnection charges are also set by the Commission. When complex issues will take months or years to resolve, interim rules keep the planning process going. The Commission's philosophy is that utilities and independent power producers have an unequal relationship and that careful rules are needed to encourage a competitive new industry.[87]

In other parts of the United States, public policies governing electric utilities and independent power producers have taken different forms in recent years. Virtually all states now require utilities to pay small-scale power producers for the electricity they generate. But some allow utilities to negotiate each contract individually, a process that can drag on for years and discourages many potential energy producers. Many contracts allow power prices to vary quarterly or monthly, a tenuous arrangement that makes it difficult for independent producers to obtain bank loans because the amount of revenue available for repayment is uncertain. Negotiated power prices range from 1¢ per kilowatt-hour to 8¢ per kilowatt-hour, making small-scale power generation infeasible in some regions and lucrative in others.[88]

In many states utilities only pay for the avoided cost of fuel, not for the avoided cost of building new power plants that would otherwise have been needed. This makes sense if the alternative source cannot provide steady power when electricity needs are greatest. But based on data gathered so far, most small-scale power sources deserve at least partial payment for the generating capacity they provide. Although the power output of some small-scale power sources such as wind turbines fluctuates rapidly, other sources, such as cogeneration and geothermal plants, are steady producers. Utilities with considerable coal- or nuclear-fired capacity have generally been more reluctant to encourage the development of new power sources, and often do not offer capacity credits. Coal and nuclear plants usually represent large capital commitments, and decommissioning or running them intermittently is much less economical than with oil- or gas-fired plants. As a result, change will come more slowly.

When a range of small-scale power sources is spread over a wide area, aggregate reliability may exceed that of large nuclear plants. For example, it is not uncommon for two 1,000-megawatt nuclear plants in the same region to be shut down simultaneously. Utility planners must be prepared to supply that much electricity from another source, even if the shutdowns occur during peak power demand. By contrast, 2,000 megawatts of small-scale power capacity might involve a hundred generating facilities using four different technologies spread over hundreds of square kilometers. Some of that capacity is likely to be out of service at any given time due to technical problems or weather conditions, but all of it will never be shut down at once. The overall performance of small-scale power sources might well approach or exceed the 55 percent of maximum output now averaged by U.S. nuclear plants.[89]

The value of a particular power source to a utility system is difficult to calculate. How a given technology affects the reliability of the whole system must be assessed, as well as the cost of building hypothetical plants in the future. Utility planners tend to discount the reliability of power sources they do not directly control, and often use simplistic analytic techniques that lead to low avoided cost payments. Many

"Utility planners tend to discount the reliability of power sources they do not directly control."

planners argue that when excess capacity exists, as is frequently the case now, the extra capacity provided by small-scale power sources is superfluous and should not be compensated. Meanwhile, some of these same utilities have plans for expensive plants to meet anticipated future growth.

The competitive struggle between utilities and independent power producers has turned bitter in many areas. In New York the Consolidated Edison Company has fought to exclude potential competitors at regulatory hearings and in the courts, on what some cogenerators view as unreasonable and even illegal grounds. The utility has lost most of these battles but has greatly slowed the development of small-scale power generation in New York. James Bruce, the frustrated chairman of the Idaho Power Company wonders, "How can I supply electricity to 265,000 customers when I don't know how many entrepreneurs will be operating next year?" Similar tensions and conflicts exist in California, but a spirit of compromise pushed by the state government has allowed the new energy projects to flourish anyway.[90]

Texas presents an interesting contrast to California. Metropolitan Houston is home to many cogeneration projects that, coupled with others, could supply much of the area's power. But the region also has excess capacity, and Houston Lighting and Power has so far paid only avoided fuel costs for the electricity it buys. In 1984, under the watchful eye of the state utility commission, the utility signed cogeneration contracts with avoided capacity payments for the first time. Houston Lighting and Power still plans to build two coal-fired plants likely to cost substantially more than cogeneration. Meanwhile, potential cogenerators are expected to bid against one another for contracts. This Texas-sized struggle pits some of the largest oil and chemical companies against a giant utility. But without established procedures or a clear strategy by either the utility or the state commission, the struggle remains largely unproductive. Without new policy initiatives, the area's cogeneration potential will never be fully harnessed.[91]

Most utility commissions in the sixties and seventies did little more than rubber stamp the decisions of industry executives. Now they must mediate complicated and contentious disputes between utilities and independent energy producers. Some utility commissions are clearly in over their heads, but procedures established by California, Montana, North Carolina, and a few other states provide models that others can follow.[92] The trend is toward higher avoided cost prices and standard contracts that ease negotiations. But much more can be done. Independent power producers must be permitted to compete fairly with conventional electricity projects and become part of the mainstream of utility planning. If today's excess capacity results in neglect of these issues, new decentralized power sources will not be available when they are needed.

Improved energy efficiency presents a different set of challenges to existing utility systems. Although many companies now sponsor conservation programs, few of these approach their potential, and rarely are construction programs and efficiency investments compared equally. To reduce electricity use to its cost-effective level, market forces must be put to work. Utilities can play a crucial role in bringing this about.

One of the few efforts to fully promote residential energy efficiency began in Hood River, Oregon, in 1983, sponsored by Pacific Power and Light and the Bonneville Power Administration. Prompted by studies by the Natural Resources Defense Council (NRDC), contractors are installing up to $4,000-worth of free conservation measures—such as efficient water heaters, triple-glazed windows, and extra thick insulation—in each of several thousand homes. By charging nothing, the utility has persuaded 60 percent of its customers to sign up for the program.[93]

Preliminary results show the program cutting power use by more than half at a cost to the utility of only 3¢ per kilowatt-hour, far less than the cost of new power sources. According to David Goldstein of NRDC, this project demonstrates that most other utility conservation

"Many utility executives still regard conservation as a public relations effort."

efforts are relatively unproductive "cream skimmers" that miss many cost-effective measures. They also make it harder for owners of some partially retrofitted homes to justify full weatherization in the future. Goldstein believes that efficiency's full potential will be realized only when the range of options is professionally assessed, utilities pay the full cost, and regulators include the investment in the utilities' rate base.[94]

The Hood River project is part of the Northwest Conservation and Electric Power Plan, established under Congressional mandate in 1983 in response to cost overruns and the eventual cancellation of four nuclear plants in Washington State. The Northwest Plan, a product of epic political battles over the energy future of the region, is one of the first efforts by a utility system—composed of both private and public companies—to treat efficiency and new generating capacity equally. Conservation is specifically required where it is the more cost-effective approach. The goal is to achieve almost 5,000 megawatts-worth of conservation by the end of the century. By reducing projected load in the year 2000 from 27,000 megawatts to 22,000 megawatts, the plan will allow the region to avoid any new plant construction until 1998.[95]

Conservation programs in most areas are proceeding at a crawl when compared to their potential. Many utility executives still regard conservation as a public relations effort to impress regulators and politicians, rather than an integral part of utility strategy. Recently some executives have tried to abandon the limited conservation programs they do have, arguing that slower demand growth and excess capacity make them unnecessary. Other utilities have revived their "marketing" programs, again encouraging customers to use more power and soak up excess capacity. Abandoning conservation and promoting power use may make short-term profits for some utilities and temporarily restrain electricity rates, but these strategies will be costly and counterproductive in the long run. Regulators should require energy efficiency programs, unless utility executives are willing to risk their own funds trying to build new plants at an equivalent cost.[96]

Some utility planners argue that regardless of conservation's cost-effectiveness, the rate at which consumers will buy more efficient electric motors or insulate their homes is unpredictable and only slightly delays the "hard decision" to build additional power plants. But analysts can now predict fairly confidently the outcome of a given conservation program, as well as its economic merits. In fact, well-managed conservation programs should reduce the largest uncertainty facing utility planners today: the rate of demand growth. If utilities invest in efficiency and install conservation measures, demand forecasts will be more reliable. By investing in electricity conservation, utilities in effect purchase a reduction in uncertainty, bringing stability to power planning.[97]

With utilities investing in efficiency and entrepreneurs building small-scale generating plants, the traditional boundaries of the power business are rapidly breaking down. All indications are that these boundaries will continue to crumble in the coming years. The energy services industry expands yearly and is finding vast potential for improved energy efficiency. It is time planners considered seriously the possibility of slowly falling power use, something that would undermine many economic assumptions on which this industry rests.

Utility planners looking for new capacity are themselves increasingly considering modular units such as efficient combined-cycle combustion turbines or small-scale fluidized-bed coal plants. Even many nuclear engineers are now convinced that nuclear power can only be revived with small-scale modular plants that have fail-safe features. The era of the 1,000-megawatt-plus thermal power plant is coming to an end, and utilities and independent energy producers are in a sense competing to lead the way in modular power generation. So far, the independents are winning.[98]

The many changes in the power industry raise fundamental questions that go well beyond the tinkering with traditional electricity policies that has occurred so far. Power generation may no longer be a

"Power generation may no longer be a natural monopoly."

natural monopoly. Although transmission and distribution of power are most effectively done by a large government-owned or -regulated company, smaller competitive companies can probably better develop new technologies and build new generating plants. The small-scale power phenomenon has already resulted in *de facto* deregulation of power generation in some areas. Taken a step further, utilities could be prohibited from plant construction, opening up competition among private companies in building plants and selling power to customers. Eventually, even existing central power plants could be operated independently of utilities. Earnings would be based on how efficiently the plants are run.[99]

Under such a system, electric utilities would be "common carriers" similar to pipelines or railroads that link producers and customers. Utilities would help forecast and plan, and channel funds to customers for improved efficiency. Governments would set rates as well as efficiency and environmental standards. To prevent coal plants from harming the environment or human health, pollution emission standards would still be needed even in a "deregulated" industry, and, of course, governments would still have to set nuclear safety standards. The small-scale power industry already has similar controls. Cogeneration plants must meet government air pollution standards, and environmental impact statements are often required for wind farms and geothermal projects.

The utility system as it was organized in most countries in the early part of the century simply cannot meet today's challenges and opportunities. But exactly how it should be structured in the future is uncertain. Some utility executives argue that the industry could be made lean and competitive through the development of deregulated subsidiaries. In the United States, 104 utilities are conducting wind power research projects and 56 have solar power projects. Many are considering plans to build commercial plants in the future. Taking another approach, Wisconsin Power and Light has entered the wind turbine business, and Alabama Power is building a photovoltaics manufacturing plant.[100]

Few utilities are ready to make such a bold leap into the competitive world unless forced by regulators or changing circumstances. Both the structure and history of the industry discourage rapid change, and most executives still seem preoccupied with fighting innovation rather than harnessing it. The rules of the game need to be changed. If the electric industry is opened up, the utilities themselves might become more innovative.

The world's electricity systems have barely changed, but already the potential for a major transition in the decades ahead is evident. Since 1980 orders for central power plants have greatly slowed in many countries, and in the interim, a surprising array of alternative strategies has emerged. With the right incentives, opportunities for improving electricity efficiency and using decentralized technologies are enormous. It may be possible to forego not only oil- and gas-powered generation in many areas, but also coal-fired plants, which are among the heaviest contributors to the world's most pressing pollution problems.

More fundamental changes may be ahead. David Morris of the Institute for Local Self-Reliance believes that today's new generating sources are only a prelude to the most revolutionary of technologies—photovoltaic cells which if placed on rooftops could make each house its own power plant. Peter Hunt, a Virginia-based energy consultant, has a similar vision.[101] He believes that within a decade both photovoltaics and fuel cells will fall in cost to the point where homeowners will call up the local utility and "tell them to come get the damned meter," completely disconnecting from the electricity grid.

Such a scenario is now possible and perhaps even likely in some regions. But while some independent producers are disconnecting from the grid, long-distance transfer of electricity will likely increase to take advantage of huge differences in generating costs between regions. Already Canada is becoming a major power exporter to the United States, and northern and southern Europe are making similar transfers. Electricity grids will make it possible for independent producers to "wheel" their power hundreds of miles to consumers.

The future will likely bring a combination of large utility grids, smaller "mini grids," and many independent households and industries. Though complicated, such a system could be easily run and monitored by computer. A mixed system would also reduce overall costs and yet allow many users to operate independently if the wider grid shut down. Massive blackouts such as the one that hit much of the eastern United States in 1964 might become a thing of the past.

Technological change and institutional reform of the electricity system are now reinforcing themselves, and the long-run results may surprise even the most visionary thinkers. Whether complete decentralization ever occurs, moving in this direction is the best way to contain electricity costs and improve the industry's environmental record.

CHRISTOPHER FLAVIN is a Senior Researcher with Worldwatch Institute and coauthor of *Renewable Energy: The Power to Choose* (W. W. Norton, Spring 1983). His research deals with renewable energy technologies and policies. He is a graduate of Williams College, where he studied Economics and Biology and participated in the Environmental Studies Program.

1. Thomas P. Hughes, *Networks of Power: Electrification in Western Society, 1880-1930* (Baltimore, Md: The Johns Hopkins University Press, 1983).

2. "Electric Power," *Encyclopedia Britannica*, 15th Edition, 1976.

3. David Morris, *Be Your Own Power Company* (Emmaus, Penn.: Rodale Press, 1983).

4. Martin G. Glaeser, *Public Utilities in American Capitalism* (New York: Macmillan Publishing Co., 1957); Marquis Childs, *The Farmer Takes a Hand: The Electric Power Revolution in Rural America* (Garden City, N.Y.: Doubleday Co., 1952).

5. Thomas Hughes, *Networks of Power*.

6. "Electric Power," *Encyclopedia Britannica*.

7. *Ibid.*

8. *Ibid.*

9. The World Bank, *The Energy Transition in Developing Countries* (Washington, D.C.: 1983).

10. The World Bank, *The Energy Transition in Developing Countries;* Howard S. Geller, "The Potential for Electricity Conservation in Brazil," Companhia Energetica de São Paulo, February 1984, unpublished.

11. Robert Sadove, "Economics and World Bank Financing of Coal-Based Electric Power Projects in Developing Countries," *Natural Resources Forum*, Vol. 8, No. 1, 1984.

12. The World Bank, *The Energy Transition in Developing Countries;* Hugh Collier, *Developing Electric Power: Thirty Years of World Bank Experience* (Baltimore, Md.: The Johns Hopkins University Press, 1984); "Record Lending for World Bank," *The Energy Daily*, August 1, 1984.

13. U.S. Energy Information Administration, *Thermal Electric Plant Construction Cost and Annual Production Expenses, 1980* (Washington, D.C.: 1983); U.S. Energy Information Administration, *1983 Annual Energy Review* (Washington, D.C.: 1984).

14. Energy Information Administration, *1983 Annual Energy Review.*

15. Scott Fenn, *America's Electric Utilities: Under Siege and in Transition* (New York: Praeger Publishers, 1984).

16. Richard E. Morgan, "Federal Tax Expenditures for Electric Utilities," testimony before the U.S. House of Representatives Committee on Ways and Means, July 21, 1983.

17. Federal Energy Regulatory Commission, "Report of Cost and Quality of Fuels for Electric Plants," January 1984, unpublished.

18. Energy Information Administration, *Thermal Electric Plant Construction Cost.*

19. U.S. Congress, Office of Technology Assessment, *Acid Rain and Transported Air Pollutants* (Washington, D.C.: 1984).

20. Charles Komanoff, *Power Plant Cost Escalation: Nuclear and Coal Capital Costs, Regulation and Economics* (New York: Komanoff Energy Associates, 1981).

21. T.A. Burnet et al., "Economic Evaluation of Limestone and Lime Flue Gas Desulfurization Processes," U.S. Environmental Protection Agency research paper, March 1984; H.A. Cavanaugh, "Utility Cleanup Spending to Drop 23%," *Electrical World*, July 1984.

22. Carlos Murawczyk and Ken M. Moy, "Environmental Protection from Power Generation: An International Overview," *Public Utilities Fortnightly*, April 28, 1983; Environmental Resources Limited, *Acid Rain: A Review of the Phenomenon in the EEC and Europe* (Brussels: Graham & Trotman, 1983).

23. International Energy Agency, *World Energy Outlook* (Paris: Organisation for Economic Co-operation and Development, 1982); "Nuclear: World Status," *Financial Times Energy Economist*, January 1983.

24. Christopher Flavin, *Nuclear Power: The Market Test* (Washington, D.C.: Worldwatch Institute, December 1983).

25. U.S. Energy Information Administration, *Monthly Energy Review* (Washington, D.C.: June 1984); United Nations, *Annual Review of Electric Energy*

Statistics for Europe (New York: June 1984); Haruki Tsuchiya, "Energy Future for Japan," draft paper for the Research Institute for Systems Technology, Tokyo, undated.

26. U.S. Energy Information Administration, *Electric Power Monthly*, (Washington, D.C.: April 1984).

27. U.S. Energy Information Administration, *Monthly Energy Review*, June 1984; United Nations, *Annual Bulletin of Electric Energy Statistics for Europe*, various years.

28. "34th Annual Electric Utility Industry Forecast," *Electrical World*, September 1983.

29. Irwin Stelzer and David Roe, "Viewpoint," *Electrical World*, May 1982.

30. Fenn, *America's Electric Utilities*.

31. John McCaughey, "Loanshark Financing: Troubled Utilities Discover That Money Isn't Cheap," *Energy Daily*, August 20, 1984.

32. Author's calculation based on data compiled in Cambridge Energy Research Associates, *Prometheus Bound: Nuclear Power at the Turning Point* (Cambridge, Mass.: 1983).

33. "For Utilities, 1983 was a Very Good Year," *Energy Daily*, April 17, 1984; Alan J. Nogee, "Rate Shock: Confronting the Cost of Nuclear Power," Environmental Action Foundation, Washington, D.C., October 1984.

34. Mark Clifford, "Utilities: After the Calm, the Storm," *Financial World*, June 13-26, 1984.

35. Matthew L. Wald, "Utilities' Chapter 11 Prospects," *The New York Times*, June 26, 1984.

36. "France: Nuclear Over-Capacity Even Before 1985," *European Energy Report*, July 1983.

37. "Forecasting the Patterns of Demand," *EPRI Journal*, December 1982.

38. North American Electric Reliability Council, *Electric Power Supply & Demand* (Princeton, N.J.: 1983).

39. Assumes an average nuclear construction cost of $2,500 per kilowatt (1983 dollars) based on actual costs of U.S. nuclear plants being completed in the mid-eighties.

40. Worldwatch Institute estimates based on various sources.

41. "34th Annual Electric Utility Industry Forecast"; U.S. Energy Information Administration, *Annual Energy Outlook* (Washington, D.C.: 1984).

42. C.C. Burwell, "The Role of Electricity in American Industry," research brief for the Institute for Energy Analysis, Oak Ridge, Tennessee, June 1984; Peter Navarro, "Our Stake in the Electric Utility's Dilemma," *Harvard Business Review*, May-June, 1982; John R. Siegel and John O. Sillin, "Changes in the Real Price of Electricity: Implications for Higher Load Growth," *Public Utilities Fortnightly*, September 15, 1983.

43. Craig R. Johnson, "Why Electric Power Growth Will Not Resume," *Public Utilities Fortnightly*, April 14, 1983; Solar Energy Research Institute, *A New Prosperity: Building a Renewable Energy Future* (Andover, Mass.: Brick House Publishing, 1981).

44. Amory B. Lovins and L. Hunter Lovins, testimony before the Subcommittee on Energy Conservation and Power, Committee on Energy and Commerce, U.S. House of Representatives, February 7, 1984.

45. U.S. Department of Energy, *The Future of Electric Power in America: Economic Supply for Economic Growth* (Washington, D.C.: 1983).

46. Harold B. Finger, "The Growing Importance of Electricity: Early Warnings of a Developing Crisis," presented to the 11th Energy Technology Conference, March 1984; Eugene N. Oatman, "If the Lights Go Out? — The Day After," presented to the 11th Energy Technology Conference, Washington, D.C., March 1984.

47. U.S. Congress, *Public Utility Regulatory Policies Act of 1978* (Washington, D.C.: 1978); Federal Energy Regulatory Commission, *Small Power Production and Cogeneration Facilities; Regulations Implementing Section 210 of the Public Utility Regulatory Policies Act of 1978* (Washington, D.C.: 1980).

48. Lehman Brothers Kuhn Loeb Research, "Cogeneration: State of the Art in Electric Utility Development," August 25, 1982; U.S. Congress Office of Technology Assessment, *Industrial and Commercial Cogeneration* (Washington, D.C.: 1983); Glenn H. Lovin, "The Resurgence of Cogeneration in the United States," paper presented to the New York Society of Security Analysts, April 4, 1984.

49. Marc H. Ross and Robert H. Williams, *Our Energy: Regaining Control* (New York: McGraw-Hill, 1981).

50. Joseph A. Glorioso, "Cogeneration: A Technology Reborn," *Industry Week*, January 23, 1984.

51. R.L. Walzel, "Dow Experience with Cogeneration," paper presented to the Executive Conference on Cogeneration, date and location unknown; "Big New Cogeneration Plant for Houston," *Energy Daily*, September 9, 1983; "Dow Chemical, Houston Utility at Loggerheads Over Power," *Energy Daily*, August 27, 1984; J.R. Bickman, Houston Lighting and Power, private communication, October 22, 1984.

52. Joseph A. Glorioso, "Cogeneration: A Technology Reborn."

53. Ravi K. Sakhuja, "Modular Cogeneration for Commercial Light Industrial Sector," *Cogeneration World*, January/February 1984; Frost and Sullivan study cited in Stuart Diamond, "Cogeneration Jars the Power Industry," *The New York Times*, June 10, 1984.

54. The Sievert Group, "Packaged Gas Fired Cogeneration Systems for Fast Food Restaurants," paper presented at 11th Energy Technology Conference, Washington, D.C., March 1984; Paul Johnson, "McDonald's Looks At Cogeneration," *Diesel Progress*, July 1984.

55. Peter Hunt, Peter Hunt Associates, private communication, October 31, 1984.

56. Office of Technology Assessment, *Industrial and Commercial Cogeneration*.

57. Resource Planning Associates, Inc., *"The Potential for Industrial Cogeneration Development by 1990,"* (Cambridge, Mass.: 1981); U.S. Department of Energy, *Industrial Cogeneration Potential: Targeting of Opportunities at the Plant Site* (Washington, D.C.: 1983).

58. Frost and Sullivan study cited in Stuart Diamond, "Cogeneration Jars the Power Industry."

59. Mike Batham, California Energy Commission, private communication, September 27, 1984.

60. Mike Batham, California Energy Commission, private communication, September 27, 1984; Mike Ringer, "Relative Costs of Electricity Production," California Energy Commission staff report, October 1984.

61. Christopher Flavin, *Wind Power: A Turning Point* (Washington, D.C.: Worldwatch Institute, July 1981); Wind farm proposals in various states and nations are author's compilation based on review of industry literature.

62. Pacific Gas and Electric Company, "The Geysers Power Plant Development," internal memorandum, March 26, 1982; Ronald DiPippo, "Development of Geothermal Electric Power Production Overseas," paper presented at the 11th Energy Technology Conference, Washington, D.C., March 1984.

63. Ronald DiPippo, "Development of Geothermal Electric Power."

64. William A. Loeb, "How Small Hydro is Growing Big," *Technology Review*, August/September 1983; U.S. use and cost figures from Raymond J. O'Connor, chairman of the Federal Energy Regulatory Commission, response to inquiry by the Subcommittee on Energy Conservation and Power, House Committee on Energy and Commerce, February 17, 1984.

65. Government of Sweden, "Green Power: Biofuels Are a Growing Concern," *Scientific American*, August 1984; James L. Easterly and Elizabeth C. Saris, "A Survey of the Use of Biomass as a Fuel to Produce Electric Energy in the United States," paper presented at the 11th Energy Technology Conference, Washington, D.C., March 1984.

66. Hal Mitchell, "Current Day Biomass Technologies," and Robert P. Kennel, "Ultrapower: An Idea Whose Time Has Come," papers presented at the Renewable Energy Technologies Symposium and International Exhibition, Anaheim, California, August 1984.

67. Dairy farmer quoted in Richard Munson, *The Power Makers*, Rodale Press, forthcoming.

68. "Philippines Produces Wood Power," *World Solar Markets*, August 1983.

69. Edgar A. DeMeo and Roger W. Taylor, "Solar Photovoltaic Power Systems: An Electric Utility R&D Perspective," *Science*, April 20, 1984; "Fuel Cells for the Nineties," *EPRI Journal*, September 1984.

70. "Progress and Tradition in Energy Conservation," *Chikyu no Koe*, November 1981; Howard Geller, "Residential Appliances and Space Conditioning Equipment: Current Savings Potential, Cost Effectiveness and Residential Needs," paper presented at the American Council for an Energy Efficient Economy 1984 Summer Study on Energy Efficiency in Buildings, Santa Cruz, California, June 1984.

71. Howard Geller, *Energy Efficient Appliances* (Washington, D.C.: American Council for an Energy-Efficient Economy, 1983).

72. "Evolution in Lighting," *EPRI Journal*, June 1984.

73. "Evolution in Lighting," *EPRI Journal*, June 1984; David B. Goldstein, "Wasted Light: An Economic Rationale for Saving 75% of Lighting Energy in Commercial Buildings," 1984, unpublished.

74. Lee Schipper, "Residential Energy Use in the OECD: 1970-1982," and Lee Schipper, "Energy-Efficient Housing in Sweden," papers presented at the American Council for an Energy Efficient Economy 1984 Summer Study on Energy Efficiency in Buildings, Santa Cruz, California, June 1984.

75. Marc Ross, "Industrial Energy Conservation," *Natural Resources Journal*, April 1984; "Electricity: Lever on Industrial Productivity," *EPRI Journal*, October 1984.

76. "Pacing Plant Motors for Energy Savings," *EPRI Journal*, March 1984; Walter J. Martiny, "Making the Choice Between Normal & Hi-Efficiency Motors," paper presented at the 11th Energy Technology Conference, Washington D.C., March 1984.

77. Solar Energy Research Institute, *A New Prosperity: Building a Renewable Energy Future* (Andover, Mass.: Brick House Publishing, 1981); Amory B. Lovins and L. Hunter Lovins, testimony before the Subcommittee on Energy

Conservation and Power, Committee on Energy and Commerce, U.S. House of Representatives, February 7, 1984.

78. Sarah Glazer, "The Residential Conservation Service: Expectations, Performance and Potential for the Future," *Energy Conservation Bulletin*, July-August 1984; Alliance to Save Energy, *Utility Promotion of Investment in Energy Efficiency: Engineering, Legal, and Economic Analyses* (Washington D.C.: 1984).

79. Douglas Cogan and Susan Williams, *Generating Energy Alternatives* (Washington, D.C.: Investor Responsibility Research Center, 1983); Dorothy Dickey, Mark D. Levine, and James E. McMahon, "Effects of Utility Incentive Programs for Appliances on the Energy Efficiency of Newly Purchased Appliances," paper presented at the American Council for an Energy Efficient Economy 1984 Summer Study on Energy Efficiency in Buildings, Santa Cruz, California, June 1984.

80. M. Centaro, Florida Power & Light, private communication, October 11, 1984.

81. Cogan and Williams, *Generating Energy Alternatives*; Bill Smith and Kathy Litelle, Carolina Power & Light, private communication, October 11, 1984; Lee Callaway, Pacific Gas and Electric Company, private communication, October 11, 1984.

82. Frederick Pickle, New England Electric, private communication, October 11, 1984.

83. Larry Condelli et al., "Improving Utility Conservation Programs: Outcomes, Interventions, and Evaluations," paper presented at the American Council for an Energy Efficient Economy 1984 Summer Study on Energy Efficiency in Buildings, Santa Cruz, California, June 1984.

84. Ralph Mitchell quoted in "Why Utilities Should Get Into the Conservation Business," *Energy Daily*, October 29, 1984; Michael Foley quoted in Matthew L. Wald, "Adding Power But No Plants," *The New York Times*, July 6, 1984.

85. U.S. Department of Energy, *The Future of Electric Power in America*.

86. Author's estimate based on data in California Energy Commission, *Secur-*

ing California's Energy Future: 1983 Biennial Report (Sacramento, Calif.: 1983), Pacific Gas and Electric Company, "Long-Term Planning Results 1984-2004," May 1984, and William R. Ahern, "Policy Report on Resource Options for Southern California Edison Company: General Rate Case for Test Year 1985," California Public Utilities Commission, April 1984; Jan Hamrin, private communication, November 2, 1984.

87. William R. Ahern, California Public Utilities Commission, private communication, May 30, 1984.

88. N. Richard Friedman, Resource Dynamics Corporation, "State Rulemaking and Utility Pricing for Cogeneration: National Trends in PURPA Implementation," paper presented at the Renewable Energy Technologies Symposium and International Exhibition, Anaheim, California, August 1983; David K. Owens, Edison Electric Institute, "Overview of the States Regulations and Rate Settings Under PURPA Section 210," paper presented at the Renewable Energy Technologies Symposium and International Exhibition, Anaheim, California, June 1984.

89. Robert D. Morris, "The Electrical Energy Production System in Transition: The Critical Factor of Reliability," chapter in Howard J. Brown, edit., *Decentralizing Electricity Production* (New Haven, Conn.: Yale University Press, 1983).

90. Consolidated Edison's disputes with small-scale power producers are described in Stuart Diamond, "Do It Yourself Electricity on the Rise," *The New York Times*, June 24, 1984. James Bruce quoted in Richard Munson, *The Power Makers*, Emmaus, Penn., Rodale Press, forthcoming.

91. "Dow Chemical, Houston Utility at Loggerheads Over Power," *Energy Daily*, August 27, 1984; Public Utility Commission of Tex., *Cogeneration and Small Power Production in Texas* (Austin, Texas: 1983).

92. The various approaches to PURPA implementation and the implications for independent energy producers are described in David Morris, *Be Your Own Power Company* (Emmaus, Penn.: Rodale Press, 1983).

93. Bonneville Power Administration, "BPA Launches the Hood River Conservation Project," November 1983; H. Gil Peach, Terry Oliver, and David B. Goldstein, "Cooperation & Diversity in a Large-Scale Conservation Research

Project," paper presented at the American Council for an Energy Efficient Economy 1984 Summer Study on Energy Efficiency in Buildings, Santa Cruz, California, June 1984.

94. David Goldstein, Natural Resources Defense Council, private communication, May 30, 1984.

95. Northwest Power Planning Council, *Northwest Conservation and Electric Power Plan* (Portland, Ore.: 1983).

96. "Utilities Gear Up for New Marketing Thrust," *Electrical World*, August 1982; "Utilities are Tempting Big Customers to Turn Up the Juice," *Business Week*, October 31, 1983.

97. Ralph Cavanagh, "Electrical Energy Futures, *Environmental Law*, Vol. IV:133.

98. "New Capacity in Smaller Packages," *EPRI Journal*, May 1983; "Future Plants: What Kind Will They Be?", *Electrical World*, July 1984; Lawrence M. Lidsky, "The Reactor of the Future?" *Technology Review*, February/March 1984.

99. One of the first proposals for electric utility deregulation was John Bryson, "Electric Utilities: the Next Ten Years," California Public Utilities Commission, March 27, 1981; also see David A. Huettner, "Restructuring the Electric Utility Industry: A Modest Proposal," in Brown, edit., *Decentralizing Electricity Production*.

100. John Bryson, Southern California Edison Company, private communication, June 7, 1984.

101. David Morris talk at the Renewable Energy Technologies Symposium and International Exhibition, Anaheim, California, June 1984; Peter Hunt, private communication, July 19, 1984.

THE WORLDWATCH PAPER SERIES

No. of Copies

______ 1. **The Other Energy Crisis: Firewood** by Erik Eckholm.
______ 2. **The Politics and Responsibility of the North American Breadbasket** by Lester R. Brown.
______ 3. **Women in Politics: A Global Review** by Kathleen Newland.
______ 4. **Energy: The Case for Conservation** by Denis Hayes.
______ 5. **Twenty-two Dimensions of the Population Problem** by Lester R. Brown, Patricia L. McGrath, and Bruce Stokes.
______ 6. **Nuclear Power: The Fifth Horseman** by Denis Hayes.
______ 7. **The Unfinished Assignment: Equal Education for Women** by Patricia L. McGrath.
______ 8. **World Population Trends: Signs of Hope, Signs of Stress** by Lester R. Brown.
______ 9. **The Two Faces of Malnutrition** by Erik Eckholm and Frank Record.
______ 10. **Health: The Family Planning Factor** by Erik Eckholm and Kathleen Newland.
______ 11. **Energy: The Solar Prospect** by Denis Hayes.
______ 12. **Filling The Family Planning Gap** by Bruce Stokes.
______ 13. **Spreading Deserts—The Hand of Man** by Erik Eckholm and Lester R. Brown.
______ 14. **Redefining National Security** by Lester R. Brown.
______ 15. **Energy for Development: Third World Options** by Denis Hayes.
______ 16. **Women and Population Growth: Choice Beyond Childbearing** by Kathleen Newland.
______ 17. **Local Responses to Global Problems: A Key to Meeting Basic Human Needs** by Bruce Stokes.
______ 18. **Cutting Tobacco's Toll** by Erik Eckholm.
______ 19. **The Solar Energy Timetable** by Denis Hayes.
______ 20. **The Global Economic Prospect: New Sources of Economic Stress** by Lester R. Brown.
______ 21. **Soft Technologies, Hard Choices** by Colin Norman.
______ 22. **Disappearing Species: The Social Challenge** by Erik Eckholm.
______ 23. **Repairs, Reuse, Recycling—First Steps Toward a Sustainable Society** by Denis Hayes.
______ 24. **The Worldwide Loss of Cropland** by Lester R. Brown.
______ 25. **Worker Participation—Productivity and the Quality of Work Life** by Bruce Stokes.
______ 26. **Planting for the Future: Forestry for Human Needs** by Erik Eckholm.
______ 27. **Pollution: The Neglected Dimensions** by Denis Hayes.
______ 28. **Global Employment and Economic Justice: The Policy Challenge** by Kathleen Newland.
______ 29. **Resource Trends and Population Policy: A Time for Reassessment** by Lester R. Brown.
______ 30. **The Dispossessed of the Earth: Land Reform and Sustainable Development** by Erik Eckholm.
______ 31. **Knowledge and Power: The Global Research and Development Budget** by Colin Norman.
______ 32. **The Future of the Automobile in an Oil-Short World** by Lester R. Brown, Christopher Flavin, and Colin Norman.
______ 33. **International Migration: The Search for Work** by Kathleen Newland.
______ 34. **Inflation: The Rising Cost of Living on a Small Planet** by Robert Fuller.
______ 35. **Food or Fuel: New Competition for the World's Cropland** by Lester R. Brown.
______ 36. **The Future of Synthetic Materials: The Petroleum Connection** by Christopher Flavin.
______ 37. **Women, Men, and The Division of Labor** by Kathleen Newland.

_____ 38. **City Limits: Emerging Constraints on Urban Growth** by Kathleen Newland.

_____ 39. **Microelectronics at Work: Productivity and Jobs in the World Economy** by Colin Norman.

_____ **n Potential**

_____ mith.

_____ **ment**

_____ el Deudney.

_____ **s**

_____ en Newland.

_____ n and Pamela

_____ n Newland.

_____ udney.

_____ **Power**

_____ **s**

_____ er R. Brown.

_____ **nall Families**

_____ iel Deudney

_____ U. Chandler

_____ andra Postel

_____ illiam U.

_____ ester Brown and

_____ **ll-Scale**

Single Copy—$4.00

Bulk Copies (any combination of titles)
2-5: $3.00 each 5-20: $2.00 each 21 or more: $1.00 each

Calendar Year Subscription (1984 subscription begins with Paper 58)
U.S. $25.00 _____

Make check payable to Worldwatch Institute
1776 Massachusetts Avenue NW, Washington, D.C. 20036 USA

Enclosed is my check for U.S. $ _____

name

address

city **state** **zip/country**